BLAST OFF to 6th Grade Math

5th to 6th Grade Transition Math Workbook

This book belongs to:

____

____

____

GET THE ANSWER KEY

The answer key for this book is available as a free PDF download.

SCAN HERE

ISBN: 978-1-64824-042-3
EDventure Learning LLC
5601 State Route 31 #1296
Clay, NY 13039

www.edventurelearning.com

Email us at hello@edventurelearning.com
Follow us on social media @edventurelearning

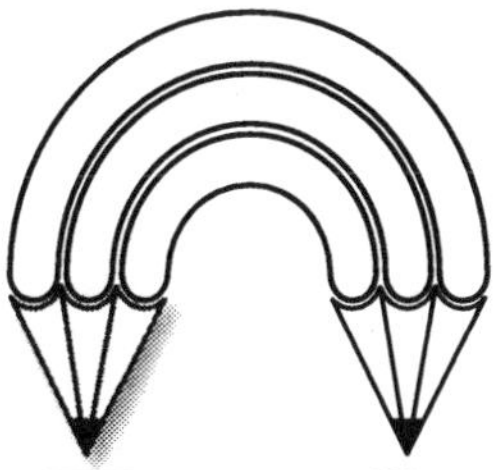

5 - 6

This book is part of the
Bridge Builder Learning Series,
designed to help bridge the gap between grade levels. The content in this book is designed to review math from 5th grade and set students up for success in 6th grade.

CONTENTS

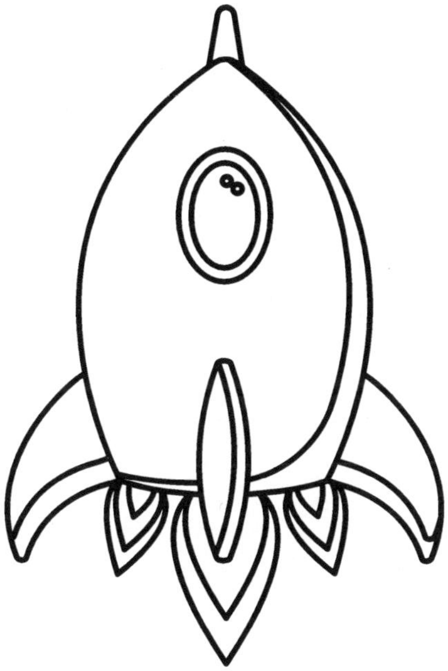

Numbers and Place Value

6 Place Value

Write the value of the underlined digit in each number.

1. 6,0**9**3,215 ______________
2. 4,**7**28,016 ______________
3. **8**,301,479 ______________
4. 1,92**4**,436 ______________
5. 2,503,17**5** ______________
6. **7**85,932 ______________
7. 9,748,6**1**9 ______________
8. 5**6**3,074 ______________
9. 9**3**,856 ______________
10. 3,091,**2**87 ______________

Circle the number in each pair in which the digit **8** has a **greater** value.

11. 10,298 8,245
12. 2,816,492 7,982,563
13. 465,892 392,087
14. 5,238,175 5,285,731
15. 63,508 55,681
16. 8,347,126 876,072

Fill in the blanks with the correct numbers.

17. In 7,298,564, the digit ______ is in the hundred thousands place.
18. In 6,581,047, the digit ______ is in the millions place.
19. In 4,921,068, the digit ______ is in the ten thousands place.
20. In 8,397,450, the digit ______ is in the hundreds place.

Place Value

Fill in the correct bubble to answer each question.

1. Which of these numbers has an even digit in the tens place?

   - ○ 865,735
   - ○ 641,023
   - ○ 547,892

2. Which of these numbers has a thousands digit greater than 4?

   - ○ 1,025,981
   - ○ 5,941,630
   - ○ 6,084,273

3. Which of these numbers has a zero in the ten thousands place?

   - ○ 1,080,760
   - ○ 8,920,003
   - ○ 7,200,406

4. Which of these numbers has a digit whose value is 300,000?

   - ○ 3,456,207
   - ○ 4,315,980
   - ○ 2,637,941

5. Which of these numbers has an odd digit in the millions place?

   - ○ 3,489,127
   - ○ 2,187,095
   - ○ 4,183,647

6. Which of these numbers has a ten thousands digit less than 5?

   - ○ 9,761,548
   - ○ 549,032
   - ○ 72,381

8 Place Value

Write each number in both standard and expanded form.

(EX) Two million, four hundred thirty-five thousand, two hundred ten

Standard form: 2,435,210

Expanded form: 2,000,000 + 400,000 + 30,000 + 5,000 + 200 + 10

(1) One million, two hundred thirteen thousand, four hundred fifteen

Standard form: ______________________

Expanded form: ______________________

____

(2) Nine hundred fifty-seven thousand, six hundred forty-eight

Standard form: ______________________

Expanded form: ______________________

____

(3) Six million, twenty-three thousand, five hundred eighty-seven

Standard form: ______________________

Expanded form: ______________________

____

Place Value

Fill in the blanks with the correct numbers.

(EX) 6 thousands = ___60___ hundreds
= ___600___ tens

(1) 5 ten thousands = __________ hundreds
= __________ tens

(2) 7 hundreds = __________ tens
= __________ ones

(3) 2 hundred thousands = __________ thousands
= __________ ones

(4) 4 millions = __________ hundred thousands
= __________ tens

(5) 9,000 ones = __________ thousands
= __________ tens

(6) 30 tens = __________ hundreds
= __________ ones

10 Rounding and Estimation

Fill in the chart to round each number to the given place value.

3,157,624	
Ten	3,157,620
Hundred	3,157,600
Thousand	3,158,000
Ten Thousand	3,160,000
Hundred Thousand	3,200,000
Million	3,000,000

2,683,057	
Ten	
Hundred	
Thousand	
Ten Thousand	
Hundred Thousand	
Million	

4,394,248	
Ten	
Hundred	
Thousand	
Ten Thousand	
Hundred Thousand	
Million	

6,285,103	
Ten	
Hundred	
Thousand	
Ten Thousand	
Hundred Thousand	
Million	

5,721,635	
Ten	
Hundred	
Thousand	
Ten Thousand	
Hundred Thousand	
Million	

8,164,652	
Ten	
Hundred	
Thousand	
Ten Thousand	
Hundred Thousand	
Million	

Rounding and Estimation

Write a whole number that fits each description. There will be more than one possible correct answer.

1. A number that rounded to the nearest ten equals 8,760

   ____

2. A number that rounded to the nearest hundred equals 1,000

   ____

3. An even number that rounded to the nearest ten equals 790

   ____

4. An odd number that rounded to the nearest ten equals 300

   ____

5. A number that rounded to the nearest hundred equals 18,500 and rounded to the nearest thousand equals 18,000

   ____

12 Rounding and Estimation

Fill in the correct bubble to answer each question.

1. The city of Springfield has a population of 113,394, and the surrounding suburbs have a population of 92,119. Which of these is the best estimate of the total population of Springfield and its suburbs?

   ○ 20,000 ○ 120,000 ○ 200,000

2. A dairy cow produces an average of 6 gallons of milk a day. Which is the best estimate of the number of gallons of milk that a farm with 287 dairy cows can expect to produce in a day?

   ○ 1,800 ○ 8,000 ○ 12,000

3. 483 candies are divided into 8 bags. Which of these is the best estimate for the number of candies in each bag?

   ○ 40 ○ 50 ○ 60

4. Roger is driving from New York to California, a trip that will be 2,794 miles long. On the first day, he drove 492 miles. Which of these is the best estimate for how many miles Roger has left on his trip?

   ○ 2,000 ○ 2,300 ○ 3,300

Rounding and Estimation

Use the information below to answer the questions that follow.

Leslie and Charles are shopping for new living room furniture. The prices for some pieces of furniture they are considering are shown in the table at right.

Item	Price
Standard Sofa	$1,529
Love Seat	$1,209
Sectional Sofa	$3,179
Recliner	$499
Coffee Table	$399

1. Leslie wants to buy a standard sofa and a love seat. Charles wants to buy a sectional sofa instead. Leslie argues that it would be about $700 less expensive to buy the standard sofa and love seat rather than the sectional. Is she correct? Explain.

____

____

____

2. If Charles and Leslie purchased the standard sofa, love seat, recliner, and coffee table, what would be their approximate total cost (to the nearest hundred)?

____

14 Powers of Ten

Rewrite each expression as a power of ten.

EX $10 \times 10 \times 10 =$ 10^3

1. $10 \times 10 \times 10 \times 10 \times 10 =$ ________

2. $10 \times 10 \times 10 \times 10 =$ ________

3. $10 \times 10 =$ ________

4. $10 \times 10 \times 10 \times 10 \times 10 \times 10 \times 10 \times 10 =$ ________

5. $10 \times 10 \times 10 \times 10 \times 10 \times 10 \times 10 \times 10 \times 10 \times 10 =$ ________

Rewrite each exponent as a multiplication expression.

EX $10^3 =$ $10 \times 10 \times 10$

6. $10^7 =$ ________________________________

7. $10^1 =$ ________________________________

8. $10^6 =$ ________________________________

9. $10^9 =$ ________________________________

10. $10^{12} =$ ________________________________

Powers of Ten

Evaluate each exponent.

(EX) $10^4 =$ 10,000

(1) $10^2 =$ ______

(2) $10^3 =$ ______

(3) $10^6 =$ ______

(4) $10^7 =$ ______

(5) $10^1 =$ ______

(6) $10^8 =$ ______

(7) $10^5 =$ ______

(8) $10^{10} =$ ______

(9) $10^9 =$ ______

Rewrite each number as a power of ten.

(EX) 10,000 = 10^4

(10) 100,000 = ______

(11) 100,000,000 = ______

(12) 1,000 = ______

(13) 1,000,000,000 = ______

(14) 100 = ______

(15) 10 = ______

(16) 10,000,000 = ______

(17) 1,000,000 = ______

(18) 10,000,000,000 = ______

Powers of Ten

Evaluate each expression.

(EX) $3 \times 10^2 =$ 300

(1) $5 \times 10^3 =$ ______

(2) $6 \times 10^1 =$ ______

(3) $9 \times 10^4 =$ ______

(4) $2 \times 10^5 =$ ______

(5) $4 \times 10^6 =$ ______

(6) $12 \times 10^2 =$ ______

(7) $37 \times 10^3 =$ ______

(8) $21 \times 10^1 =$ ______

(9) $50 \times 10^4 =$ ______

Complete each equation with the correct power of ten.

(EX) $2{,}500 = 25 \times 10^2$

(10) $30{,}000 = 3 \times$ ______

(11) $800 = 8 \times$ ______

(12) $970 = 97 \times$ ______

(13) $700{,}000 = 7 \times$ ______

(14) $11{,}000 = 11 \times$ ______

(15) $8{,}900{,}000 = 89 \times$ ______

(16) $60{,}000 = 60 \times$ ______

(17) $260{,}000 = 26 \times$ ______

(18) $400{,}000 = 4 \times$ ______

Powers of Ten

Fill in the correct bubble to answer each question.

1. Which of these is equivalent to 5×10^4?

   ○ 5,000 ○ 50,000 ○ 500,000

2. Which of these is equivalent to 63×10^3?

   ○ 630 ○ 6,300 ○ 63,000

3. Which of these is equivalent to 70×10^2?

   ○ 700 ○ 7,000 ○ 70,000

4. Which of these is equivalent to 870?

   ○ 87×10^1 ○ 87×10^2 ○ 87×10^3

5. Which of these is equivalent to 6,000,000?

   ○ 6×10^6 ○ 6×10^7 ○ 60×10^4

6. Which of these is equivalent to 350,000?

   ○ 35×10^3 ○ 35×10^4 ○ 35×10^5

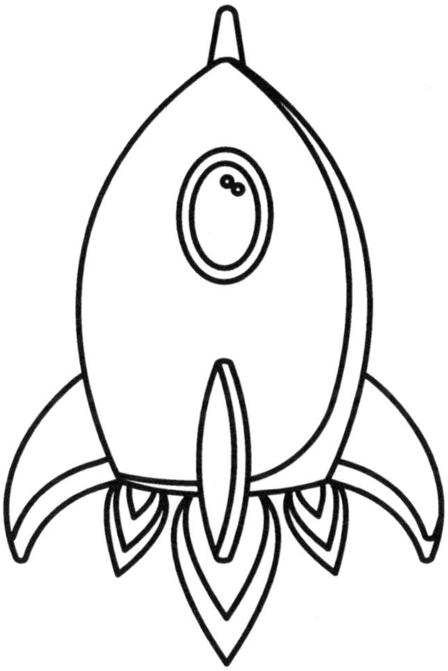

Multiplication and Division

20 Factors and Multiples

List all the factors of each number.

1. 12: ______________________________
2. 45: ______________________________
3. 60: ______________________________
4. 28: ______________________________
5. 39: ______________________________
6. 24: ______________________________

Color the factors of 36 red. Color the factors of 48 blue. If a number is a factor of both 36 and 48, color it purple.

1	2	3	4	5	6	7	8
9	10	11	12	13	14	15	16
17	18	19	20	21	22	23	24
25	26	27	28	29	30	31	32
33	34	35	36	37	38	39	40
41	42	43	44	45	46	47	48

Factors and Multiples

Find the greatest common factor (GCF) of each pair of numbers.

① 36 and 56

Factors of 36: ______________________________

Factors of 56: ______________________________

GCF of 36 and 56: ________

② 27 and 54

Factors of 27: ______________________________

Factors of 54: ______________________________

GCF of 27 and 54: ________

③ 45 and 81

Factors of 45: ______________________________

Factors of 81: ______________________________

GCF of 45 and 81: ________

④ GCF of 26 and 64: ______

⑤ GCF of 42 and 54: ______

⑥ GCF of 75 and 90: ______

⑦ GCF of 18 and 72: ______

22 Factors and Multiples

Color in all of the prime numbers on the hundreds chart below.

1	2	3	4	5	6	7	8	9	10
11	12	13	14	15	16	17	18	19	20
21	22	23	24	25	26	27	28	29	30
31	32	33	34	35	36	37	38	39	40
41	42	43	44	45	46	47	48	49	50
51	52	53	54	55	56	57	58	59	60
61	62	63	64	65	66	67	68	69	70
71	72	73	74	75	76	77	78	79	80
81	82	83	84	85	86	87	88	89	90
91	92	93	94	95	96	97	98	99	100

1. Explain how you know if a number is prime.

____

____

2. What is a non-prime number called? ____________________

Complete the factor tree to find the prime factorization of each number.

(EX)

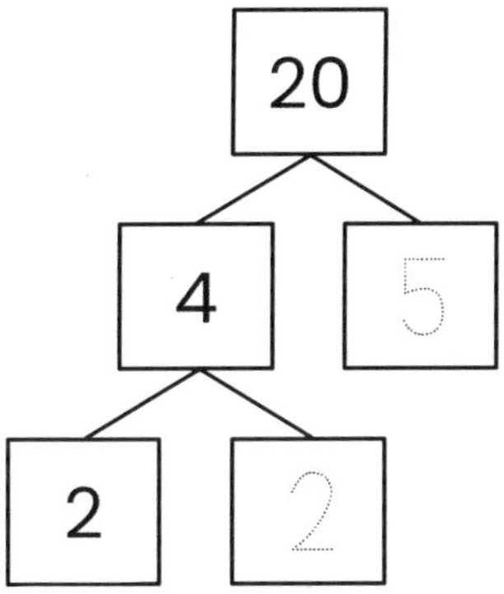

Prime factorization of 20:

2 × 2 × 5

(1)

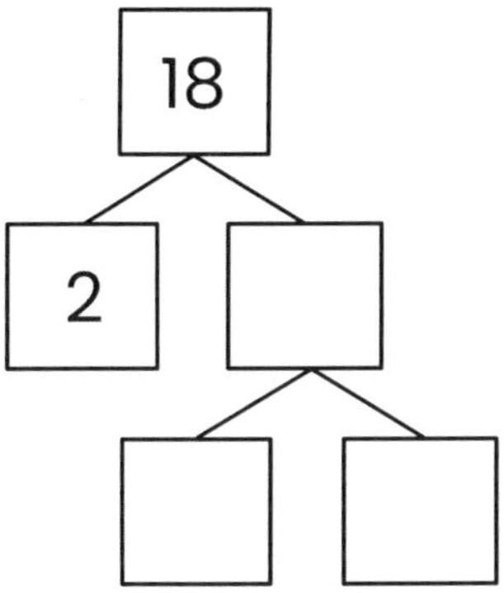

Prime factorization of 18:

____

(2)

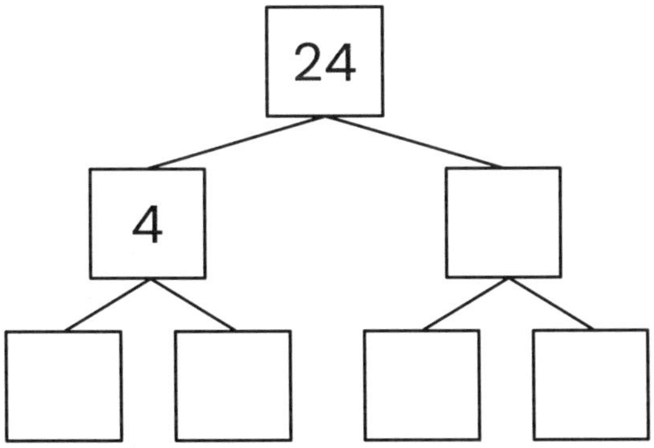

Prime factorization of 24:

____

(3)

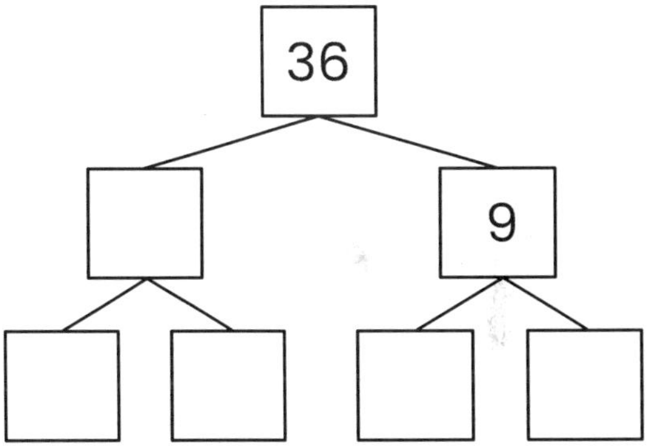

Prime factorization of 36:

____

(4)

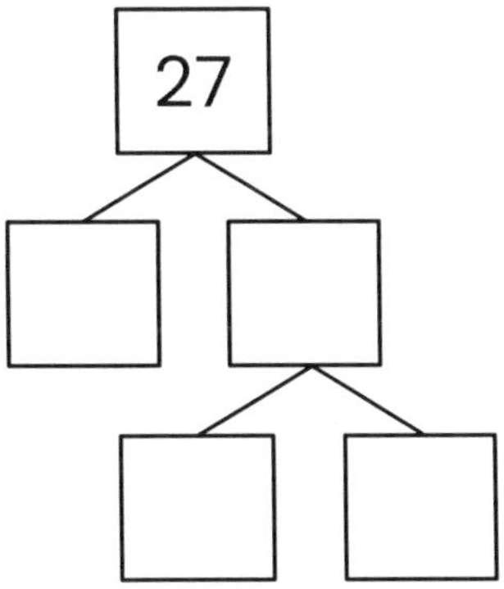

Prime factorization of 27:

____

(5)

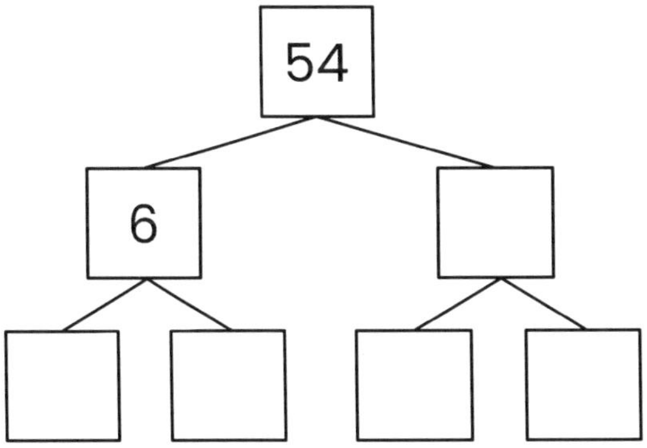

Prime factorization of 24:

____

24 Factors and Multiples

Draw a factor tree to find the prime factorization of each number.

16 Prime factorization of 16: ____________________	30 Prime factorization of 30: ____________________
48 Prime factorization of 48: ____________________	64 Prime factorization of 64: ____________________

Factors and Multiples

List the first five multiples of each number given.

1. 6: ______________________________
2. 12: ______________________________
3. 7: ______________________________
4. 11: ______________________________
5. 20: ______________________________
6. 15: ______________________________

Color the multiples of 3 red. Color the multiples of 4 blue. If a number is a multiple of both 3 and 4, color it purple.

1	2	3	4	5	6	7	8	9	10
11	12	13	14	15	16	17	18	19	20
21	22	23	24	25	26	27	28	29	30
31	32	33	34	35	36	37	38	39	40
41	42	43	44	45	46	47	48	49	50

26 Factors and Multiples

Find the least common multiple (LCM) of each pair of numbers.

1. 6 and 14

   Multiples of 6: ____________________

   Multiples of 14: ____________________

   LCM of 6 and 14: ________

2. 10 and 12

   Multiples of 10: ____________________

   Multiples of 12: ____________________

   LCM of 10 and 12: ________

3. 8 and 10

   Multiples of 8: ____________________

   Multiples of 10: ____________________

   LCM of 8 and 10: ________

4. LCM of 8 and 15: ______

5. LCM of 9 and 12: ______

6. LCM of 6 and 16: ______

7. LCM of 4 and 6: ______

Factors and Multiples

Find the number that answers each riddle and write it in the box.

1. I am an even number that is not composite.
   What number am I?

2. My prime factorization is $2 \times 3 \times 3 \times 5$.
   What number am I?

3. I am a prime number between 20 and 30.
   My digits add up to a prime number greater than 10.
   What number am I?

4. I am a composite number between 30 and 40.
   All of my factors are even except 1.
   What number am I?

5. I am a two-digit number less than 20.
   My LCM with 8 is 56.
   What number am I?

28 Multiplication

Next to each equation, write the multiplication property being shown: **associative**, **commutative**, **distributive**, **identity**, or **zero**.

1. $21 \times 7 = 7 \times 21$ ____________________

2. $371 \times 1 = 371$ ____________________

3. $(14 \times 6) \times 5 = 14 \times (6 \times 5)$ ____________________

4. $89 \times 0 = 0$ ____________________

5. $5 \times (8 + 9) = (5 \times 8) + (5 \times 9)$ ____________________

Draw an area model to find each product.

EX $362 \times 4 =$ 1,448

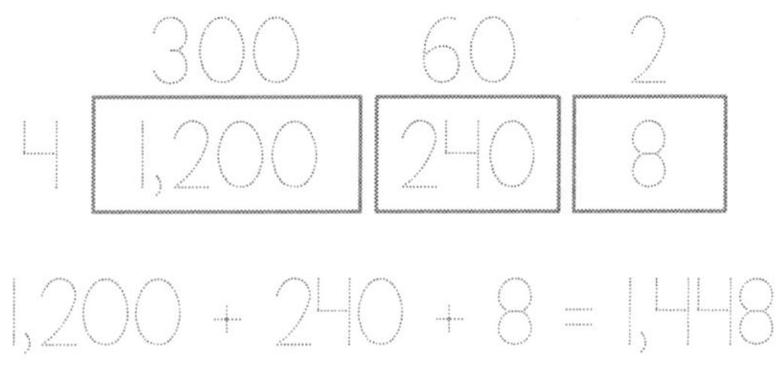

6. $981 \times 3 =$ __________

7. $79 \times 8 =$ __________

8. $1{,}427 \times 5 =$ __________

Multiplication

Use the distributive property to find each product.

EX 98 × 3 = (90 × 3) + (8 × 3) = 270 + 24 = 294

1. 47 × 6 = ______
2. 62 × 8 = ______
3. 113 × 4 = ______
4. 703 × 2 = ______
5. 398 × 7 = ______

Find each product using the standard algorithm.

42 × 6	87 × 2	53 × 9	98 × 7	35 × 4
175 × 3	436 × 8	690 × 2	502 × 6	214 × 5
2,841 × 4	6,734 × 7	8,315 × 9	7,089 × 5	3,856 × 3

30 Multiplication

Find each product using the standard algorithm.

$\begin{array}{r} 52 \\ \times\ 63 \\ \hline 156 \\ +3{,}120 \\ \hline 3{,}276 \end{array}$	$\begin{array}{r} 84 \\ \times\ 21 \\ \hline \end{array}$	$\begin{array}{r} 19 \\ \times\ 54 \\ \hline \end{array}$	$\begin{array}{r} 36 \\ \times\ 90 \\ \hline \end{array}$	$\begin{array}{r} 70 \\ \times\ 78 \\ \hline \end{array}$
$\begin{array}{r} 132 \\ \times\ 67 \\ \hline \end{array}$	$\begin{array}{r} 405 \\ \times\ 98 \\ \hline \end{array}$	$\begin{array}{r} 716 \\ \times\ 20 \\ \hline \end{array}$	$\begin{array}{r} 289 \\ \times\ 31 \\ \hline \end{array}$	$\begin{array}{r} 354 \\ \times\ 45 \\ \hline \end{array}$
$\begin{array}{r} 6{,}787 \\ \times\ 61 \\ \hline \end{array}$	$\begin{array}{r} 9{,}803 \\ \times\ 27 \\ \hline \end{array}$	$\begin{array}{r} 1{,}942 \\ \times\ 83 \\ \hline \end{array}$	$\begin{array}{r} 3{,}054 \\ \times\ 49 \\ \hline \end{array}$	$\begin{array}{r} 5{,}162 \\ \times\ 50 \\ \hline \end{array}$
$\begin{array}{r} 678 \\ \times\ 12 \\ \hline \end{array}$	$\begin{array}{r} 90 \\ \times\ 43 \\ \hline \end{array}$	$\begin{array}{r} 5{,}169 \\ \times\ 65 \\ \hline \end{array}$	$\begin{array}{r} 273 \\ \times\ 97 \\ \hline \end{array}$	$\begin{array}{r} 48 \\ \times\ 81 \\ \hline \end{array}$

Find each product using the standard algorithm.

192 × 384	576 × 901	283 × 475	609 × 182	374 × 560
768 15,360 +57,600 73,728				
918 × 273	645 × 190	827 × 364	519 × 280	135 × 246
810 × 791	113 × 124	517 × 682	224 × 921	683 × 235

Multiplication

Find each product.

13 × 2	46 × 5	79 × 8	81 × 9	27 × 3
465 × 20	908 × 39	172 × 48	635 × 57	491 × 61
203 × 847	495 × 560	857 × 192	610 × 837	293 × 465
2,657 × 4	4,483 × 75	6,920 × 657	1,102 × 48	3,938 × 3

Use multiplication to solve each word problem.

1. According to the nutrition label on a cereal box, each serving of cereal contains 22 grams of carbohydrates. If there are 16 servings of cereal in the box, how many total grams of carbohydrates are in the entire box?

2. It takes the planet Neptune about 165 years to orbit the Sun. How long would it take Neptune to orbit the Sun 25 times?

3. An acre is an area equal to 4,480 square yards. What is the area, in square yards, of a 400-acre farm?

4. A bakery charges $4 per loaf of bread. Last month, they sold 1,235 loaves. How much money did they make from bread sales last month?

5. Mt. Waialeale in Hawaii receives an average of 9,763 mm of rain annually. How much rain (in mm) would be expected over a 5-year period?

34 Division

Find each quotient using an area model.

(EX) 493 ÷ 6 = 82 R1

	70	+ 10	+ 2 R1
6	420	60	12

493	73	13
−420	−60	−12
73	13	1

(1) 257 ÷ 3 = ________

(2) 1,042 ÷ 8 = ________

(3) 956 ÷ 7 = ________

(4) 2,421 ÷ 9 = ________

(5) 843 ÷ 4 = ________

Division

Find each quotient using the standard algorithm.

(EX) $4\overline{)387}$ = 96 R3
-36
27
-24
3

(1) $6\overline{)1{,}750}$

(2) $5\overline{)3{,}921}$

(3) $8\overline{)2{,}017}$

(4) $3\overline{)816}$

(5) $9\overline{)647}$

(6) $7\overline{)1{,}982}$

(7) $2\overline{)7{,}302}$

(8) $4\overline{)9{,}742}$

(9) $6\overline{)762}$

(10) $9\overline{)4{,}108}$

(11) $7\overline{)21{,}564}$

Division

Find each quotient using an area model.

(EX) 2,492 ÷ 16 = 155 R12

100 + 50 + 5 R12

16 | 1,600 | 800 | 80 |

2,492 − 1,600 = 892

892 − 800 = 92

92 − 80 = 12

(1) 876 ÷ 25 = ________

(2) 5,921 ÷ 32 = ________

(3) 713 ÷ 63 = ________

(4) 3,801 ÷ 29 = ________

(5) 1,392 ÷ 81 = ________

Division

Find each quotient using the standard algorithm.

1. $23\overline{)984}$
2. $21\overline{)634}$
3. $76\overline{)853}$
4. $62\overline{)937}$
5. $18\overline{)882}$
6. $51\overline{)762}$
7. $91\overline{)927}$
8. $82\overline{)741}$
9. $47\overline{)893}$

38 Division

Find each quotient using the standard algorithm.

1. $47\overline{)3{,}809}$
2. $33\overline{)7{,}634}$
3. $13\overline{)2{,}981}$
4. $39\overline{)4{,}201}$
5. $14\overline{)2{,}945}$
6. $42\overline{)7{,}518}$
7. $98\overline{)1{,}492}$
8. $20\overline{)6{,}032}$
9. $52\overline{)6{,}821}$

Use division to solve each word problem.

1. A teacher spent $273 on books for his classroom. If each book cost $13, how many books did he buy?

2. Carmen is putting photos into an album. Each page holds four photos. How many pages will she need to hold 187 photos?

3. The five fifth grade classes at a school raised a combined $1,750 in a fundraiser. What is the average amount raised per class?

4. Allison can bake a batch of 48 mini-muffins at a time. If she needs to bake a total of 240 mini-muffins, how many batches will she need?

5. It takes Earth 24 hours to rotate once on its axis. The same process takes Venus 5,832 hours. How many rotations does Earth make in the time it takes Venus to complete one rotation?

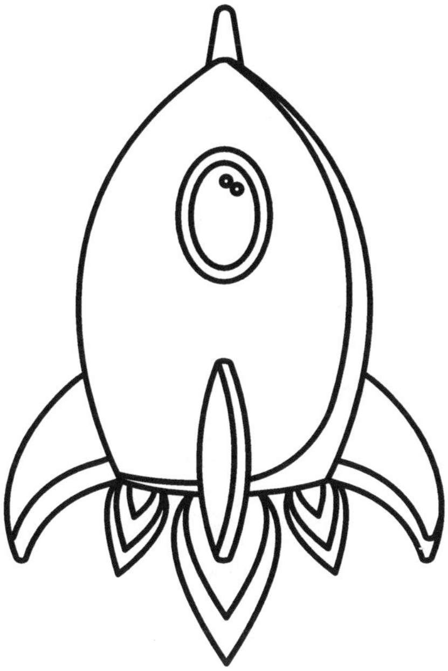

Mixed Operations

42 Number Patterns

Find the addition or subtraction rule for each sequence and fill in the missing number in each pattern.

(EX) 5, 8, 11, 14, 17

Rule: Add 3

(1) 67, 57, 47, 37, ________

Rule: ________________

(2) 54, 65, ________, 87, 98

Rule: ________________

(3) 110, ________, 94, 86, 78

Rule: ________________

(4) 331, 324, 317, ________, 303

Rule: ________________

(5) ________, 184, 197, 210, 223

Rule: ________________

Write the rule shown by each input-output table.

(EX) **Rule:** Subtract 5

Input	Output
12	7
17	10
20	15

(6) **Rule:** ________________

Input	Output
9	18
25	34
37	46

(7) **Rule:** ________________

Input	Output
13	25
30	42
42	54

(8) **Rule:** ________________

Input	Output
25	19
34	28
51	45

Find the multiplication or division rule for each sequence and fill in the missing number in each pattern.

EX 2, 6, 18, 54 , 162

Rule: Multiply by 3

1. 512, 128, 32, 8, ________

Rule: ________

2. 21, 42, ________ , 168, 336

Rule: ________

3. 1,875, ________ , 75, 15, 3

Rule: ________

4. 324, 108, 36, ________ , 4

Rule: ________

5. ________ , 6, 36, 216, 1,296

Rule: ________

Write the rule shown by each input-output table.

EX **Rule:** Divide by 10

Input	Output
70	7
120	12
250	25

6. **Rule:** ________

Input	Output
7	28
8	32
12	48

7. **Rule:** ________

Input	Output
6	30
11	55
25	250

8. **Rule:** ________

Input	Output
24	12
46	23
52	26

Number Patterns

Fill in the missing numbers on each input-output table using the given rule.

(EX) **Rule:** Add 7.

Input	Output
3	10
6	13
11	18
15	22

(1) **Rule:** Subtract 12.

Input	Output
13	1
17	
26	14
	17

(2) **Rule:** Add 8.

Input	Output
5	13
	16
12	
15	23

(3) **Rule:** Multiply by 8.

Input	Output
3	
4	32
	56
9	72

(4) **Rule:** Divide by 6.

Input	Output
12	2
36	
42	7
	12

(5) **Rule:** Multiply by 11.

Input	Output
3	33
5	
7	77
	121

Answer each word problem involving number patterns.

Derek is building a block tower consisting of several rows of blocks. He puts 16 blocks on the bottom row, 13 blocks on the second row, and 10 blocks on the third row.

1. If Derek continues this pattern, how many blocks will be in the fourth row?

2. If Derek continues this pattern until he reaches a row that has only one block, how many rows will the tower have?

3. If Derek continues this pattern until he reaches a row that has only one block, how many total blocks will be in the tower?

Sonia is putting money in a jar each day to save up to buy a toy. On the first day, she puts in a nickel. On the second day, she puts in a dime. On the third day, she puts in $0.20, and on the fourth day, she puts in $0.40.

4. If Sonia continues this pattern, how much money will she put in the jar on the fifth day?

5. If Sonia continues this pattern, what it is total amount of money she will have put in the jar at the end of six days?

Word Problems

Add, subtract, multiply, or divide to solve each word problem.

1. Envelopes come in boxes of 25. Mary needs 268 envelopes to mail invitations to a large event. How many boxes of envelopes does she need?

2. Loose leaf paper comes in packs of 250 sheets. How many sheets of paper are in 12 packs?

3. Carmen bought a phone for $499 and a case for $15. What was Carmen's total cost for the phone and case?

4. A city's population was approximately 5,717,000 in 2020. This was an increase of 268,000 since 2010. What was the population of the city in 2010?

5. At an average speed of 48 miles per hour, how long would it take a car to travel 336 miles?

Word Problems

Use the information below to answer each word problem.

The table below lists the lengths of the five longest rivers in Europe.

River	Length (miles)
Volga	2,194
Danube	1,777
Ural	1,509
Dnieper	1,367
Don	1,162

1. How much longer is the Volga River than the Ural River?

2. What is the combined length of the three longest rivers?

3. The sixth longest river in Europe is the Pechora River, which is 653 miles shorter than the Danube. How long is the Pechora?

4. What is the length of the Ural River in yards?

5. What would be the length of a river that was exactly twice as long as the Dnieper River?

Word Problems

Solve each multi-step word problem.

1. Adrian has 15 marbles, Brian has 22 marbles, and Chris has 17 marbles. If they combined all their marbles and then divided them evenly among themselves, how many marbles would each boy get?

2. Kelly can type at an average speed of 65 words per minute. Suzanne can type 72 words per minute. If they each typed non-stop for 3 minutes, how many more words would Suzanne be able to type than Kelly?

3. Thalia takes her pulse and counts 31 heartbeats in 30 seconds. What is her heart rate in beats per minute?

4. Colin baked 4 batches of cookies for a party. Each batch contained 18 cookies. By the end of the party, 53 of the cookies had been eaten. How many were left over?

Solve each multi-step word problem.

1. Sadie rides her bike 5 blocks to her friend Callie's house, then they ride together 8 blocks to school. On the way home, they take the same route in reverse. How many total blocks does Sadie ride?

2. Nick is selling candy bars for a fundraiser. Each bar costs $2. His neighbor buys 3 milk chocolate bars, 2 dark chocolate bars, 1 caramel bar, and 1 peanut butter bar from Nick. What is the total cost of the neighbor's candy order?

3. Mrs. Smith had pack of 200 sheets of construction paper. She gave 4 sheets to each of the 23 students in her class. How many sheets were left in the pack?

4. In Columbus, Ohio, the lowest recorded temperature in 2023 was 12°F. The highest recorded temperature in Columbus that year was 2° less than 8 times the lowest temperature of the year. What was the highest temperature in Columbus in 2023?

50 Expressions

Write an expression for each operation described.

(EX) Subtract 4 from 12

12 − 4

(1) Multiply 6 by 3

____

(2) The sum of 9 and 4

____

(3) 25 divided by 5

____

(4) 10 deducted from 15

____

(5) The product of 2 and 13

____

(6) 36 divided by 4

____

(7) 6 added to 15

____

(8) 5 more than ***n***

____

(9) ***x*** divided by 2

____

(10) 10 times ***m***

____

(11) ***a*** minus ***b***

____

Expressions

Write a single expression for each set of operations.

(EX) 2 subtracted from the sum of 3 and 7 ___3 + 7 - 2___

1. 7 added to the sum of 6 and 2 ________

2. Subtract 2 from 8, then subtract 1 from the result ________

3. The product of 6 and 4 divided by 8 ________

4. Divide 18 by 2, and multiply the result by 5 ________

Evaluate each expression.

(EX) $6 + 3 - 4 =$ ___5___

5. $7 - 5 + 10 =$ ________

6. $5 + 2 + 8 =$ ________

7. $18 - 2 - 6 =$ ________

8. $8 + 4 - 3 =$ ________

9. $6 - 1 + 3 =$ ________

10. $12 \div 3 \div 2 =$ ________

11. $14 \div 2 \times 3 =$ ________

12. $4 \times 6 \times 2 =$ ________

13. $5 \times 6 \div 2 =$ ________

52 Expressions

Evaluate each expression containing parentheses.

(EX) $4 \times (3 + 2) =$ 20

$4 \times 5 = 20$

(1) $(4 \times 3) + 2 =$ ______

(2) $4 + (3 \times 2) =$ ______

(3) $(4 + 3) \times 2 =$ ______

(4) $4 \times (3 - 2) =$ ______

(5) $(4 - 3) \times 2 =$ ______

(6) $4 \div (3 - 2) =$ ______

(7) $(4 \div 2) \times 3 =$ ______

(8) $(4 + 2) \div 3 =$ ______

(9) $(4 \div 2) + 3 =$ ______

Fill in the missing sign that would make each equation true.

(EX) $5 \times (6 \boxed{-} 3) = 15$

(10) $3 \times (6 \square 2) = 24$

(11) $20 - (5 \square 2) = 10$

(12) $4 + (9 \square 3) = 7$

(13) $36 \square (4 + 5) = 4$

(14) $12 \square (24 \div 2) = 24$

(15) $(16 \square 4) \times 3 = 36$

(16) $(7 \square 4) + 4 = 32$

Expressions

Write a single expression for each set of operations.

(EX) 7 times the sum of 2 and 4 $7 \times (2 + 4)$

1. 3 more than the product of 4 and 6 ____________

2. 2 less than the result of 6 times 3 ____________

3. 20 divided by the sum of 2 and 3 ____________

4. 6 multiplied by the sum of 7 and 3 ____________

5. Subtract 4 from 8, then multiply the result by 12 ____________

6. Add 4 to 18, then divide the result by 2 ____________

7. 5 more than 4 multiplied by ***p*** ____________

8. ***s*** divided by the result of 6 minus 2 ____________

54 Expressions

Evaluate each expression containing brackets and parentheses.

EX $2 \times [(10 + 2) \div 4] =$ 6

$2 \times [12 \div 4] =$

$2 \times 3 = 6$

1 $[4 + (2 \times 3)] \times 5 =$ ______

2 $24 \div [18 - (4 \times 3)] =$ ______

3 $[12 - (10 \div 2)] \times 3 =$ ______

4 $4 \times [(16 \div 2) + 1] =$ ______

5 $[3 \times (1 + 7)] - 9 =$ ______

6 $9 - [(3 + 4 + 5) \div 4] =$ ______

7 $[60 \div (6 \times 2)] + 8 =$ ______

8 $[(4 \div 2) + 9] - 5 =$ ______

9 $[(29 - 7) \div 2] \times 4 =$ ______

Use the order of operations (PEMDAS) to evaluate each expression.

EX $10^2 - 5 \times (4 + 2) =$ 70

Parentheses: $10^2 - 5 \times 6 =$

Exponent: $100 - 5 \times 6 =$

Multiply/Divide: $100 - 30 =$

Add/Subtract: 70

1. $3 + 10^3 \div (25 \times 4) =$ ____

Parentheses: ____

Exponent: ____

Multiply/Divide: ____

Add/Subtract: ____

2. $(12 \times 12) - 10^2 \div 5 =$ ____

Parentheses: ____

Exponent: ____

Multiply/Divide: ____

Add/Subtract: ____

3. $10^3 - 5 \times (24 + 16) =$ ____

Parentheses: ____

Exponent: ____

Multiply/Divide: ____

Add/Subtract: ____

Expressions

Fill in the correct bubble to answer each question.

1. Which operation should you perform **first** to evaluate the expression below?

   $3 + 20 \div (6 - 2) = ?$

   - ○ $3 + 20$
   - ○ $20 \div 6$
   - ○ $6 - 2$

2. Which operation should you perform **first** to evaluate the expression below?

   $5 + 25 \div 5 - 2 = ?$

   - ○ $5 + 25$
   - ○ $25 \div 5$
   - ○ $5 - 2$

3. Which operation should you perform **first** to evaluate the expression below?

   $(10 - 7) \times 4 + 2 = ?$

   - ○ $10 - 7$
   - ○ 7×4
   - ○ $4 + 2$

4. Which operation should you perform **first** to evaluate the expression below?

   $3 \times 4 + 12 \div 6 = ?$

   - ○ 3×4
   - ○ $4 + 2$
   - ○ $12 \div 6$

5. Which operation should you perform **first** to evaluate the expression below?

   $7 + 15 \div 3 - 1 = ?$

   - ○ $7 + 15$
   - ○ $15 \div 3$
   - ○ $3 - 1$

Evaluate each expression by following the order of operations.

1. $15 + 30 \div 5 - 2 =$ ______

2. $4 + 40 \div (6 + 2) =$ ______

3. $(15 - 6) \times 3 + 5 =$ ______

4. $8 \times 4 + 18 \div 3 =$ ______

5. $8 + 36 \div 4 - 3 =$ ______

6. $10^2 - 4 \times (9 + 3) =$ ______

7. $50 - 16 \div 4 \times 7 =$ ______

8. $5 \times 8 \div (7 + 3) =$ ______

9. $12 + 24 \div 3 - 15 =$ ______

10. $6 \times 6 + 21 \div 7 =$ ______

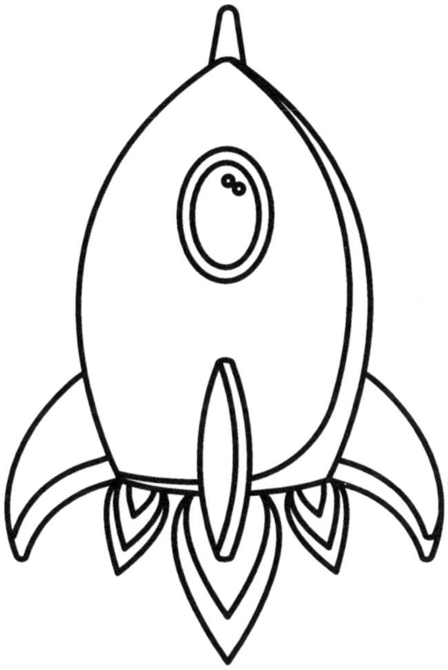

Fractions

60 Add and Subtract Fractions

Use two different colors to show the fractions on the bar model. Then, add. Write your answer in lowest terms.

EX

1

$\frac{1}{6}$	$\frac{1}{6}$	$\frac{1}{6}$	$\frac{1}{6}$	$\frac{1}{6}$	$\frac{1}{6}$

$\frac{2}{6} + \frac{1}{6} = \frac{3}{6} = \frac{1}{2}$

1

1

$\frac{1}{7}$	$\frac{1}{7}$	$\frac{1}{7}$	$\frac{1}{7}$	$\frac{1}{7}$	$\frac{1}{7}$	$\frac{1}{7}$

$\frac{2}{7} + \frac{3}{7} =$ ________

2

1

$\frac{1}{9}$	$\frac{1}{9}$	$\frac{1}{9}$	$\frac{1}{9}$	$\frac{1}{9}$	$\frac{1}{9}$	$\frac{1}{9}$	$\frac{1}{9}$	$\frac{1}{9}$

$\frac{1}{9} + \frac{5}{9} =$ ________

3

1

$\frac{1}{8}$	$\frac{1}{8}$	$\frac{1}{8}$	$\frac{1}{8}$	$\frac{1}{8}$	$\frac{1}{8}$	$\frac{1}{8}$	$\frac{1}{8}$

$\frac{3}{8} + \frac{3}{8} =$ ________

Add each pair of fractions. Write your answer in lowest terms.

4 $\frac{2}{9} + \frac{1}{9} =$ ________

5 $\frac{1}{12} + \frac{3}{12} =$ ________

6 $\frac{3}{8} + \frac{2}{8} =$ ________

7 $\frac{3}{10} + \frac{2}{10} =$ ________

8 $\frac{4}{11} + \frac{6}{11} =$ ________

9 $\frac{4}{15} + \frac{1}{15} =$ ________

10 $\frac{3}{7} + \frac{1}{7} =$ ________

11 $\frac{1}{4} + \frac{2}{4} =$ ________

12 $\frac{3}{5} + \frac{1}{5} =$ ________

13 $\frac{2}{6} + \frac{2}{6} =$ ________

Fill in the bar model to show subtraction. Then, subtract. Write your answer in lowest terms.

EX

1

$\frac{1}{6}$	$\frac{1}{6}$	$\frac{1}{6}$	$\frac{1}{6}$	$\frac{1}{6}$	$\frac{1}{6}$

$\frac{3}{6} - \frac{1}{6} = \frac{2}{6} = \frac{1}{3}$

1

1

$\frac{1}{5}$	$\frac{1}{5}$	$\frac{1}{5}$	$\frac{1}{5}$	$\frac{1}{5}$

$\frac{4}{5} - \frac{1}{5} =$ ________

2

1

$\frac{1}{9}$	$\frac{1}{9}$	$\frac{1}{9}$	$\frac{1}{9}$	$\frac{1}{9}$	$\frac{1}{9}$	$\frac{1}{9}$	$\frac{1}{9}$	$\frac{1}{9}$

$\frac{8}{9} - \frac{5}{9} =$ ________

3

1

$\frac{1}{7}$	$\frac{1}{7}$	$\frac{1}{7}$	$\frac{1}{7}$	$\frac{1}{7}$	$\frac{1}{7}$	$\frac{1}{7}$

$\frac{4}{7} - \frac{3}{7} =$ ________

Subtract each pair of fractions. Write your answer in lowest terms.

4. $\frac{5}{6} - \frac{1}{6} =$ ________

5. $\frac{4}{9} - \frac{2}{9} =$ ________

6. $\frac{7}{8} - \frac{3}{8} =$ ________

7. $\frac{9}{10} - \frac{7}{10} =$ ________

8. $\frac{11}{12} - \frac{8}{12} =$ ________

9. $\frac{13}{15} - \frac{7}{15} =$ ________

10. $\frac{5}{7} - \frac{2}{7} =$ ________

11. $\frac{3}{5} - \frac{1}{5} =$ ________

12. $\frac{10}{11} - \frac{5}{11} =$ ________

13. $\frac{9}{14} - \frac{2}{14} =$ ________

62 Add and Subtract Fractions

Use a common denominator to add or subtract each pair of fractions with unlike denominators. First, find the least common multiple (LCM) of the denominators. Re-write each fraction as an equivalent fraction with the common denominator. Then, add or subtract.

EX $\frac{2}{3} + \frac{1}{4} = ?$

LCM of 3 and 4: 12

$\frac{2}{3} = \frac{8}{12}$ $\frac{1}{4} = \frac{3}{12}$

$\frac{8}{12} + \frac{3}{12} = \frac{11}{12}$

1 $\frac{1}{5} + \frac{2}{3} = ?$

LCM of 5 and 3: ______

$\frac{1}{5} =$ ______ $\frac{2}{3} =$ ______

______ + ______ = ______

2 $\frac{7}{9} - \frac{2}{3} = ?$

LCM of 9 and 3: ______

$\frac{7}{9} =$ ______ $\frac{2}{3} =$ ______

______ - ______ = ______

3 $\frac{5}{8} - \frac{1}{6} = ?$

LCM of 8 and 6: ______

$\frac{5}{8} =$ ______ $\frac{1}{6} =$ ______

______ - ______ = ______

4 $\frac{1}{4} + \frac{3}{5} = ?$

LCM of 4 and 5: ______

$\frac{1}{4} =$ ______ $\frac{3}{5} =$ ______

______ + ______ = ______

5 $\frac{1}{2} - \frac{2}{9} = ?$

LCM of 2 and 9: ______

$\frac{1}{2} =$ ______ $\frac{2}{9} =$ ______

______ - ______ = ______

Add and Subtract Fractions

Re-write each equation so that the fractions have a common denominator. Then, add or subtract. Write the answer in lowest terms.

EX $\frac{1}{6} + \frac{3}{4} = ?$

$\frac{2}{12} + \frac{9}{12} = \frac{11}{12}$

1. $\frac{7}{10} - \frac{1}{5} = ?$

____ - ____ = ______

2. $\frac{3}{8} + \frac{5}{12} = ?$

____ + ____ = ______

3. $\frac{6}{7} - \frac{1}{2} = ?$

____ - ____ = ______

4. $\frac{1}{2} + \frac{3}{8} = ?$

____ + ____ = ______

5. $\frac{5}{8} - \frac{4}{7} = ?$

____ - ____ = ______

6. $\frac{3}{20} + \frac{3}{4} = ?$

____ + ____ = ______

7. $\frac{5}{6} - \frac{4}{9} = ?$

____ - ____ = ______

8. $\frac{1}{12} + \frac{5}{9} = ?$

____ + ____ = ______

9. $\frac{3}{7} - \frac{1}{4} = ?$

____ - ____ = ______

64 Add and Subtract Fractions

Add or subtract. Write your answer in lowest terms.

1. $\frac{3}{16} + \frac{5}{16} =$ ______

2. $\frac{17}{18} - \frac{5}{18} =$ ______

3. $\frac{2}{3} + \frac{1}{6} =$ ______

4. $\frac{11}{12} - \frac{3}{4} =$ ______

5. $\frac{1}{5} + \frac{3}{7} =$ ______

6. $\frac{3}{4} - \frac{2}{5} =$ ______

7. $\frac{3}{8} + \frac{3}{8} =$ ______

8. $\frac{5}{8} - \frac{1}{4} =$ ______

9. $\frac{7}{11} + \frac{1}{3} =$ ______

10. $\frac{9}{14} - \frac{5}{14} =$ ______

11. $\frac{2}{5} + \frac{9}{25} =$ ______

12. $\frac{7}{8} - \frac{4}{9} =$ ______

13. $\frac{5}{21} + \frac{13}{21} =$ ______

14. $\frac{8}{9} - \frac{5}{12} =$ ______

15. $\frac{5}{18} + \frac{2}{9} =$ ______

16. $\frac{23}{24} - \frac{7}{8} =$ ______

Add each set of three fractions. Write your answer in lowest terms.

1. $\frac{3}{8} + \frac{1}{4} + \frac{1}{8} =$ __________

2. $\frac{1}{6} + \frac{1}{3} + \frac{1}{2} =$ __________

3. $\frac{2}{5} + \frac{1}{10} + \frac{3}{10} =$ __________

4. $\frac{1}{3} + \frac{1}{4} + \frac{1}{6} =$ __________

5. $\frac{1}{2} + \frac{1}{4} + \frac{1}{5} =$ __________

6. $\frac{2}{9} + \frac{1}{6} + \frac{4}{9} =$ __________

7. $\frac{4}{15} + \frac{1}{3} + \frac{1}{5} =$ __________

8. $\frac{1}{7} + \frac{1}{2} + \frac{2}{7} =$ __________

9. $\frac{2}{5} + \frac{1}{6} + \frac{1}{3} =$ __________

10. $\frac{1}{6} + \frac{3}{8} + \frac{1}{4} =$ __________

66 Add and Subtract Fractions

Use the information below to answer each question.

A teacher let each student in her class choose one animal sticker. There were three different sticker designs: sloth, llama, and penguin. $\frac{1}{4}$ of the students chose a sloth sticker, $\frac{1}{3}$ chose a llama sticker, and the rest chose a penguin sticker.

1. What fraction of the students chose either a sloth or a llama?

2. What fraction of the students chose a penguin sticker?

3. Which sticker was the most popular? Explain your answer.

4. There are 24 students in the class. How many students chose a sloth sticker?

Find the sum of the mixed numbers. Write your answer in lowest terms.

(EX) $2\frac{1}{12} + 4\frac{2}{3} = \underline{6\frac{3}{4}}$

1. Add the whole numbers.

   $2 + 4 = 6$

2. Add the fractions (change to a common denominator if needed). Simplify.

   $\frac{1}{12} + \frac{2}{3} = \frac{1}{12} + \frac{8}{12} = \frac{9}{12} = \frac{3}{4}$

3. Create a mixed number.

   $6\frac{3}{4}$

(1) $5\frac{3}{7} + 3\frac{2}{7} =$ ______

(2) $1\frac{7}{10} + 4\frac{1}{5} =$ ______

(3) $2\frac{1}{3} + 6\frac{1}{2} =$ ______

(4) $7\frac{2}{11} + 2\frac{2}{9} =$ ______

(5) $4\frac{3}{16} + 3\frac{3}{8} =$ ______

(6) $3\frac{1}{3} + 5\frac{4}{9} =$ ______

68 Add and Subtract Fractions

Find the sum of the mixed numbers using regrouping. Write your answer in lowest terms.

EX $3\frac{3}{5} + 5\frac{7}{10} = 9\frac{3}{10}$

1. Add the whole numbers.

   $3 + 5 = 8$

2. Add the fractions (change to a common denominator if needed).

   $\frac{3}{5} + \frac{7}{10} = \frac{6}{10} + \frac{7}{10} = \frac{13}{10}$

3. Change the improper fraction to a mixed number. Simplify if needed.

   $\frac{13}{10} = 1\frac{3}{10}$

4. Add this to the whole number.

   $8 + 1\frac{3}{10} = 9\frac{3}{10}$

(1) $4\frac{2}{3} + 2\frac{4}{5} =$ ______

(2) $3\frac{7}{8} + 2\frac{5}{8} =$ ______

(3) $6\frac{5}{7} + 5\frac{1}{2} =$ ______

(4) $8\frac{7}{12} + 4\frac{5}{6} =$ ______

(5) $2\frac{10}{11} + 7\frac{5}{11} =$ ______

(6) $3\frac{2}{3} + 9\frac{3}{4} =$ ______

Subtract the mixed numbers. Write your answer in lowest terms.

EX $6\frac{5}{6} - 3\frac{1}{3} = 3\frac{1}{2}$

1. Change the fractions so that they have a common denominator (if needed).

$6\frac{5}{6} - 3\frac{1}{3} = 6\frac{5}{6} - 3\frac{2}{6}$

2. Subtract the whole numbers.

$6 - 3 = 3$

3. Subtract the fractions. Simplify if needed.

$\frac{5}{6} - \frac{2}{6} = \frac{3}{6} = \frac{1}{2}$

4. Create a mixed number.

$3\frac{1}{2}$

1. $7\frac{7}{12} - 5\frac{1}{4} =$ ______

2. $8\frac{9}{10} - 2\frac{4}{5} =$ ______

3. $9\frac{8}{11} - 3\frac{1}{2} =$ ______

4. $9\frac{8}{9} - 4\frac{5}{9} =$ ______

5. $4\frac{3}{4} - 2\frac{1}{8} =$ ______

6. $11\frac{14}{15} - 5\frac{3}{5} =$ ______

Add and Subtract Fractions

Find the difference between the mixed numbers using regrouping. Write your answer in lowest terms.

EX $7\frac{2}{5} - 4\frac{6}{7} = 2\frac{19}{35}$

1. Change the fractions so that they have a common denominator (if needed).

$7\frac{2}{5} - 4\frac{6}{7} = 7\frac{14}{35} - 4\frac{30}{35}$

2. Since the numerator of the first fraction is smaller than the numerator of the second fraction, regroup by borrowing from the whole number.

$7\frac{14}{35} = 6 + \frac{35}{35} + \frac{14}{35} = 6\frac{49}{35}$

3. Subtract the whole numbers from the two mixed numbers.

$6 - 4 = 2$

4. Subtract the fractions. Simplify if needed.

$\frac{49}{35} - \frac{30}{35} = \frac{19}{35}$

5. Create a mixed number.

$2\frac{19}{35}$

1. $5\frac{1}{3} - 2\frac{5}{6} =$ ______

2. $6\frac{3}{8} - 3\frac{3}{4} =$ ______

3. $8\frac{3}{11} - 6\frac{10}{11} =$ ______

4. $7\frac{1}{2} - 4\frac{7}{9} =$ ______

5. $4\frac{5}{12} - 1\frac{5}{6} =$ ______

6. $9\frac{2}{5} - 5\frac{2}{3} =$ ______

Add and Subtract Fractions

Add or subtract the mixed numbers. Regroup where needed. Write your answer in lowest terms.

1. $3\frac{1}{4} + 3\frac{3}{8} =$ ______
2. $5\frac{7}{8} - 4\frac{3}{8} =$ ______
3. $6\frac{3}{4} + 4\frac{1}{2} =$ ______
4. $7\frac{3}{5} - 1\frac{9}{10} =$ ______
5. $5\frac{7}{9} + 2\frac{5}{9} =$ ______
6. $4\frac{2}{3} - 1\frac{5}{8} =$ ______
7. $2\frac{2}{3} + 7\frac{3}{4} =$ ______
8. $9\frac{1}{6} - 5\frac{5}{12} =$ ______
9. $1\frac{5}{6} + 6\frac{2}{9} =$ ______
10. $7\frac{4}{5} - 1\frac{1}{3} =$ ______

Add and Subtract Fractions

Add or subtract. Write your answer in lowest terms.

1. $\frac{2}{5} + \frac{1}{3} =$ __________

2. $2\frac{1}{2} + 4\frac{2}{7} =$ __________

3. $\frac{9}{10} - \frac{3}{5} =$ __________

4. $7\frac{8}{11} - 3\frac{1}{3} =$ __________

5. $\frac{3}{4} + \frac{2}{3} =$ __________

6. $5\frac{3}{7} + 2\frac{2}{3} =$ __________

7. $\frac{8}{9} - \frac{5}{6} =$ __________

8. $9\frac{4}{5} - 7\frac{3}{4} =$ __________

9. $\frac{4}{7} + \frac{2}{5} =$ __________

10. $2\frac{1}{4} + 3\frac{5}{8} =$ __________

11. $\frac{11}{16} - \frac{3}{8} =$ __________

12. $8\frac{5}{9} - 4\frac{1}{3} =$ __________

13. $\frac{3}{5} + \frac{5}{8} =$ __________

14. $1\frac{3}{8} + 4\frac{1}{6} =$ __________

15. $\frac{13}{15} - \frac{4}{5} =$ __________

16. $6\frac{2}{9} - 1\frac{3}{4} =$ __________

17. $\frac{3}{7} + \frac{3}{4} =$ __________

18. $2\frac{7}{8} + 5\frac{5}{8} =$ __________

19. $\frac{9}{10} - \frac{2}{5} =$ __________

20. $5\frac{1}{4} - 3\frac{2}{5} =$ __________

Add or subtract to answer each word problem involving mixed numbers.

1. Mike bought two packages of chicken from the grocery store. One package contains $4\frac{3}{5}$ pounds of chicken and the other contains $5\frac{1}{4}$ pounds. What is the total weight of the chicken Mike purchased?

2. A zookeeper measures one of the cheetahs under her care. The head and body of the cheetah are $48\frac{1}{2}$ inches long. The cheetah's tail is $27\frac{3}{4}$ inches long. What is the total length of the cheetah from its head to the end of its tail?

3. Jack puts an empty plate onto a kitchen scale and sees that it weighs $1\frac{3}{10}$ pounds. Then, he puts a baked potato on the plate, and the scale gives a weight reading of $1\frac{4}{5}$ pounds. How much would the baked potato weigh alone?

4. To the nearest tenth of an inch, Seattle received $40\frac{9}{10}$ inches of rain in 2022 and 35 inches of rain in 2023. How many more inches of rain did Seattle receive in 2022 than in 2023?

Multiplying Fractions

Multiply each whole number by a fraction. If your answer is an improper fraction, change it to a whole or mixed number in lowest terms.

EX

$3 \times \frac{2}{3} =$ $\frac{6}{3} = 2$

1

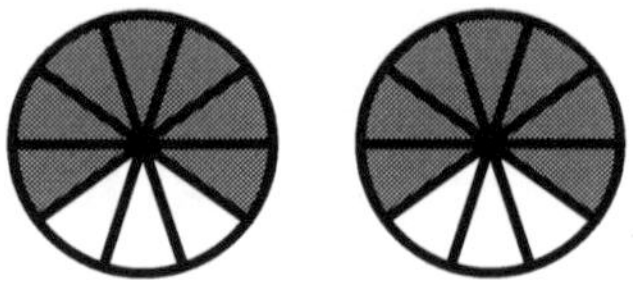

$2 \times \frac{7}{10} =$ ______

2

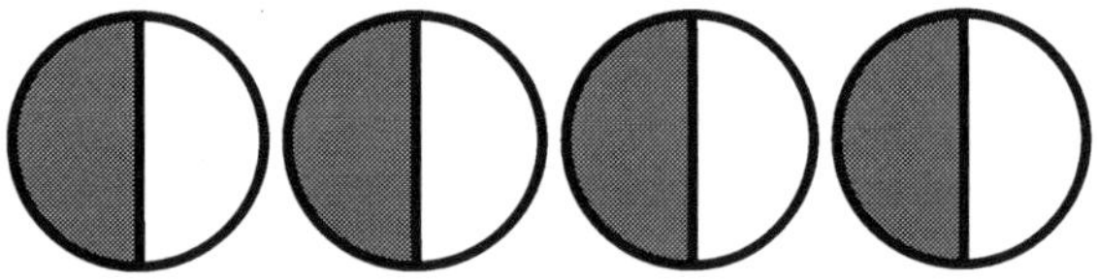

$4 \times \frac{1}{2} =$ ______

3

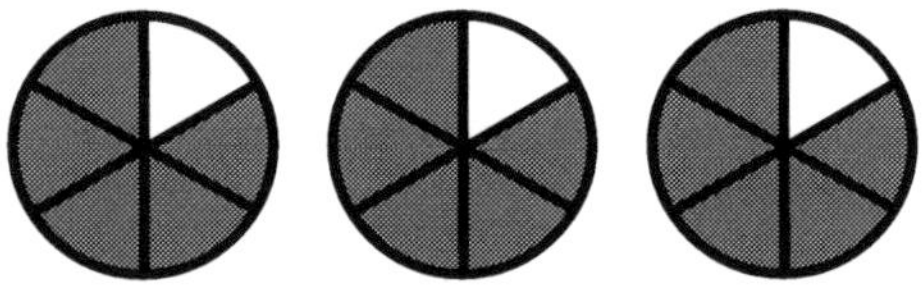

$3 \times \frac{5}{6} =$ ______

4. $5 \times \frac{2}{7} =$ ______

5. $6 \times \frac{3}{4} =$ ______

6. $4 \times \frac{3}{5} =$ ______

7. $7 \times \frac{5}{8} =$ ______

8. $8 \times \frac{4}{9} =$ ______

9. $2 \times \frac{7}{12} =$ ______

10. $5 \times \frac{3}{10} =$ ______

11. $9 \times \frac{3}{7} =$ ______

12. $6 \times \frac{3}{8} =$ ______

13. $4 \times \frac{5}{12} =$ ______

Multiplying Fractions

Multiply each pair of fractions. Write your answer in lowest terms.

EX $\frac{3}{5} \times \frac{2}{3} =$ $\frac{6}{15} = \frac{2}{5}$

1. Multiply the numerators.
2. Multiply the denominators.
3. Simplify.

(1) $\frac{1}{4} \times \frac{3}{5} =$ ________

(2) $\frac{2}{9} \times \frac{3}{4} =$ ________

(3) $\frac{4}{7} \times \frac{1}{4} =$ ________

(4) $\frac{7}{10} \times \frac{2}{5} =$ ________

(5) $\frac{5}{8} \times \frac{6}{7} =$ ________

(6) $\frac{8}{9} \times \frac{1}{2} =$ ________

(7) $\frac{9}{10} \times \frac{3}{4} =$ ________

(8) $\frac{4}{5} \times \frac{6}{11} =$ ________

(9) $\frac{5}{9} \times \frac{7}{9} =$ ________

(10) $\frac{8}{11} \times \frac{3}{5} =$ ________

(11) $\frac{2}{7} \times \frac{9}{10} =$ ________

(12) $\frac{4}{9} \times \frac{3}{7} =$ ________

(13) $\frac{1}{6} \times \frac{2}{9} =$ ________

(14) $\frac{3}{8} \times \frac{2}{3} =$ ________

76 Multiplying Fractions

Complete the area model to multiply each pair of fractions. Write your answer in lowest terms.

EX

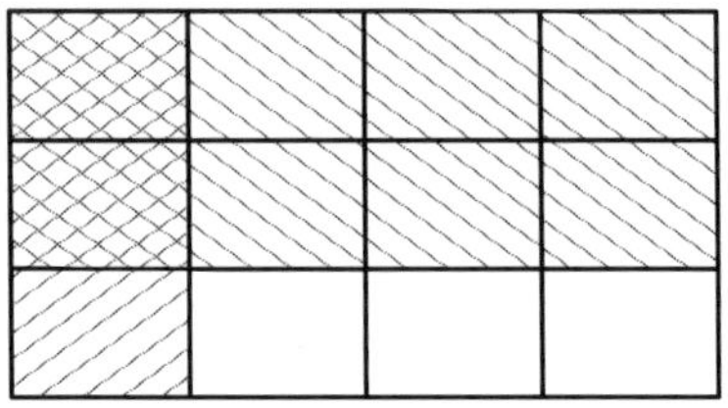

$\frac{1}{4} \times \frac{2}{3} = \frac{2}{12} = \frac{1}{6}$

1

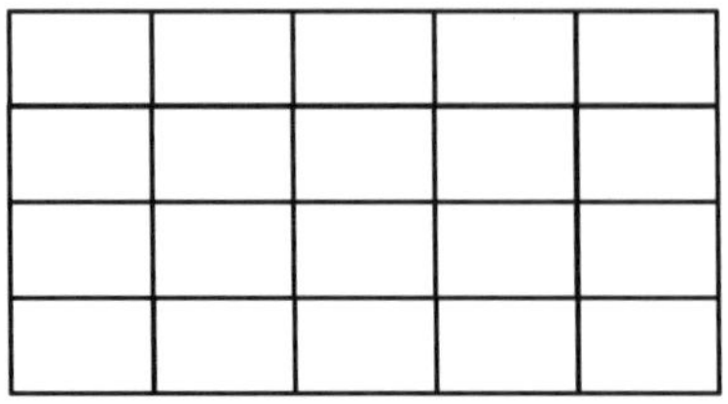

$\frac{2}{5} \times \frac{3}{4} =$ __________

2

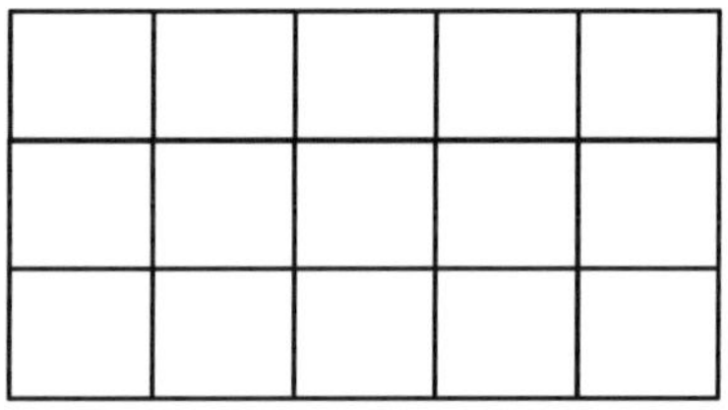

$\frac{4}{5} \times \frac{1}{3} =$ __________

3

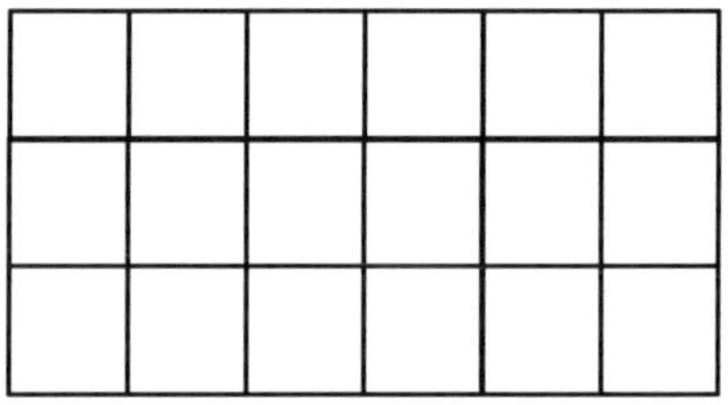

$\frac{5}{6} \times \frac{2}{3} =$ __________

4

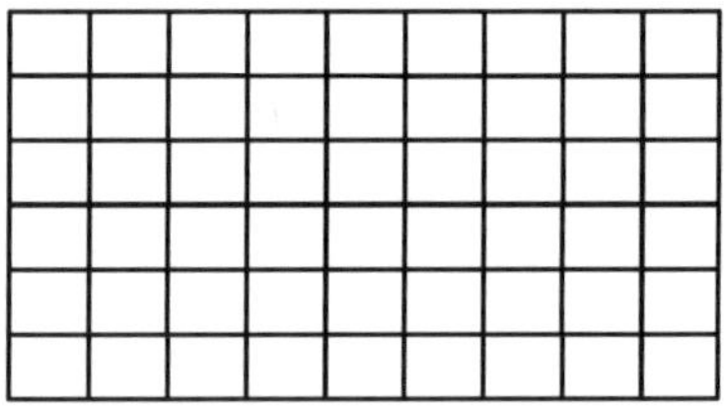

$\frac{4}{9} \times \frac{1}{6} =$ __________

5

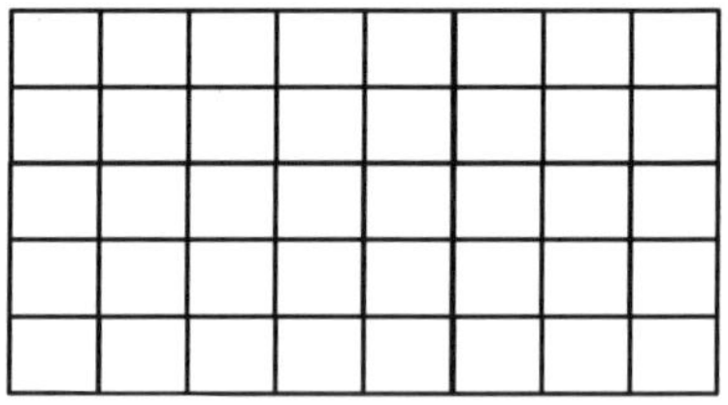

$\frac{3}{8} \times \frac{2}{5} =$ __________

6

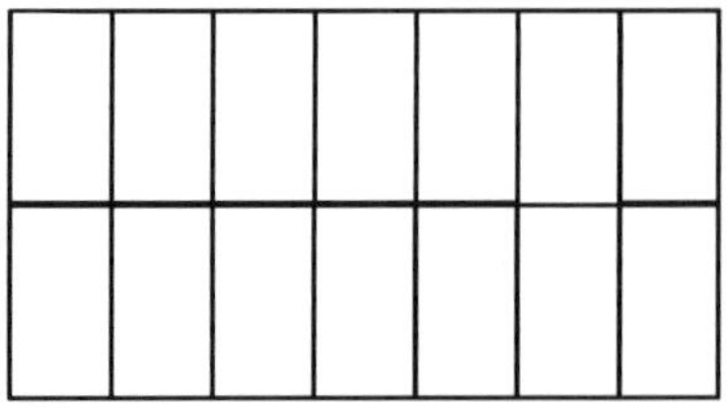

$\frac{4}{7} \times \frac{1}{2} =$ __________

7

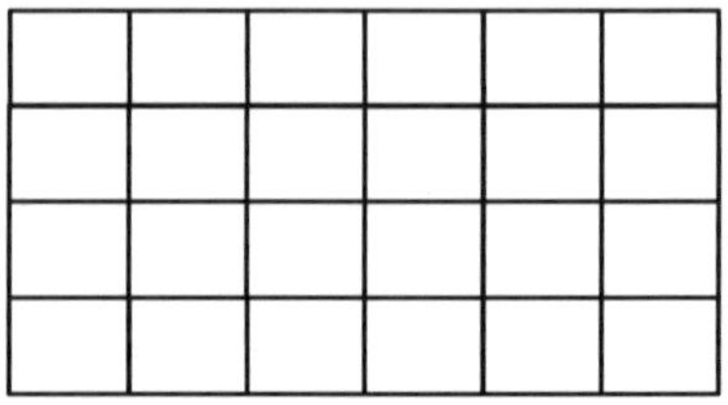

$\frac{5}{6} \times \frac{1}{4} =$ __________

Multiplying Fractions

Fill in the blanks to complete each multiplication equation.

EX $\frac{1}{5} \times \frac{4}{7} = \frac{4}{35}$

1 $\frac{2}{3} \times$ ____ $= \frac{4}{27}$

2 $\frac{5}{9} \times$ ____ $= \frac{5}{18}$

3 $\frac{3}{4} \times$ ____ $= \frac{9}{32}$

4 ____ $\times \frac{3}{2} = \frac{15}{16}$

5 ____ $\times \frac{2}{7} = \frac{2}{21}$

6 ____ $\times \frac{2}{5} = \frac{4}{15}$

7 ____ $\times \frac{7}{8} = \frac{21}{40}$

8 $\frac{1}{2} \times$ ____ $= \frac{3}{8}$

9 $\frac{4}{5} \times$ ____ $= \frac{28}{45}$

10 $\frac{7}{10} \times$ ____ $= \frac{21}{50}$

11 $\frac{5}{12} \times$ ____ $= \frac{25}{72}$

12 ____ $\times \frac{5}{7} = \frac{10}{21}$

13 ____ $\times \frac{5}{4} = \frac{15}{32}$

14 ____ $\times \frac{9}{10} = \frac{27}{70}$

15 ____ $\times \frac{13}{15} = \frac{26}{45}$

78 Multiplying Fractions

Multiply each mixed number by a whole number. Write your answer as a whole or mixed number in lowest terms.

(EX) $3\frac{2}{3} \times 5 =$ $18\frac{1}{3}$

$\frac{11}{3} \times 5 = \frac{55}{3} = 18\frac{1}{3}$

1. Change the mixed number to an improper fraction.
2. Multiply the whole number by the numerator.
3. Simplify.

(1) $4\frac{1}{4} \times 2 =$ ________

(2) $6 \times 1\frac{9}{10} =$ ________

(3) $5\frac{1}{2} \times 8 =$ ________

(4) $7 \times 2\frac{3}{5} =$ ________

Multiply each mixed number by a fraction. Write your answer in lowest terms.

(EX) $2\frac{3}{4} \times \frac{3}{8} =$ $1\frac{1}{32}$

$\frac{11}{4} \times \frac{3}{8} = \frac{33}{32} = 1\frac{1}{32}$

1. Change the mixed number to an improper fraction.
2. Multiply the numerators.
3. Multiply the denominators.
4. Simplify.

(5) $3\frac{2}{5} \times \frac{7}{9} =$ ________

(6) $\frac{5}{6} \times 4\frac{1}{3} =$ ________

(7) $1\frac{7}{8} \times \frac{2}{3} =$ ________

(8) $\frac{1}{2} \times 5\frac{2}{5} =$ ________

Multiplying Fractions

Multiply each pair of mixed numbers. Write your answer as a whole or mixed number in lowest terms.

(EX) $2\frac{1}{2} \times 4\frac{3}{4} =$ $11\frac{7}{8}$

$\frac{5}{2} \times \frac{19}{4} = \frac{95}{8} = 11\frac{7}{8}$

1. Change both mixed numbers to improper fractions.
2. Multiply the numerators.
3. Multiply the denominators.
4. Simplify.

(1) $5\frac{2}{3} \times 3\frac{1}{6} =$ ________

(2) $7\frac{1}{3} \times 2\frac{1}{4} =$ ________

(3) $1\frac{5}{8} \times 4\frac{3}{5} =$ ________

(4) $6\frac{2}{5} \times 3\frac{1}{10} =$ ________

Multiply. Write your answer as a whole or mixed number in lowest terms.

(5) $7\frac{1}{3} \times 5 =$ ________

(6) $8 \times 2\frac{7}{10} =$ ________

(7) $1\frac{2}{9} \times \frac{3}{4} =$ ________

(8) $\frac{8}{9} \times 3\frac{5}{6} =$ ________

(9) $4\frac{1}{2} \times 6\frac{1}{5} =$ ________

(10) $9\frac{1}{3} \times 1\frac{9}{10} =$ ________

80 Multiplying Fractions

Fill in the blanks with "greater than," "less than," or "equal to" to complete each sentence.

1. Multiplying any factor by 1 will result in a product that is ____________________ the first factor.

2. Multiplying any factor by a number less than 1 will result in a product that is ____________________ the first factor.

3. Multiplying any factor by a number greater than 1 will result in a product that is ____________________ the first factor.

Compare each pair of expressions using >, <, or =.

EX. $1\frac{2}{3} \times \frac{3}{4}$ (<) $1\frac{2}{3}$

4. $\frac{6}{7} \times 1$ ◯ $\frac{6}{7}$

5. $12 \times \frac{3}{2}$ ◯ 12

6. $2\frac{1}{2} \times \frac{5}{3}$ ◯ $2\frac{1}{2}$

7. $\frac{3}{5} \times 3$ ◯ $\frac{3}{5}$

8. $4 \times \frac{7}{8}$ ◯ 4

9. $5\frac{1}{8} \times \frac{4}{3}$ ◯ $5\frac{1}{8}$

10. $\frac{6}{5} \times 7$ ◯ $\frac{6}{5}$

11. $6 \times \frac{9}{10}$ ◯ 6

12. $8\frac{11}{12} \times \frac{4}{7}$ ◯ $8\frac{11}{12}$

13. $\frac{13}{15} \times 1$ ◯ $\frac{13}{15}$

14. $9 \times \frac{5}{8}$ ◯ 9

Multiplying Fractions

Multiply. Write your answer in lowest terms.

1. $4 \times \frac{2}{11} =$ ________
2. $8 \times \frac{5}{6} =$ ________
3. $\frac{7}{8} \times \frac{3}{4} =$ ________
4. $\frac{1}{2} \times \frac{4}{5} =$ ________
5. $3\frac{1}{3} \times 5 =$ ________
6. $6\frac{3}{8} \times 2 =$ ________
7. $1\frac{3}{10} \times \frac{5}{6} =$ ________
8. $2\frac{4}{5} \times \frac{2}{9} =$ ________
9. $4\frac{1}{4} \times 3\frac{2}{5} =$ ________
10. $5\frac{2}{3} \times 2\frac{3}{4} =$ ________

Multiply to solve each word problem.

11. A recipe calls for $2\frac{1}{4}$ cups of flour to make a batch of muffins. Mia wants to make 3 batches of muffins. How much flour does she need?

12. $\frac{3}{7}$ of the students in a class are boys. $\frac{2}{3}$ of the boys have brown eyes. What fraction of the total number of students in the class are boys with brown eyes?

82 Dividing Fractions

To divide with fractions, you will need to be able to find the **reciprocal** of a number. To find the reciprocal of a fraction, switch the numerator and denominator.

(EX) The reciprocal of $\frac{2}{3}$ is $\frac{3}{2}$

The reciprocal of a whole number uses 1 as the numerator.

(EX) The reciprocal of 8 is $\frac{1}{8}$

Find the reciprocal of each number.

1. The reciprocal of $\frac{3}{4}$ is ____
2. The reciprocal of 3 is ____
3. The reciprocal of $\frac{5}{6}$ is ____
4. The reciprocal of 2 is ____
5. The reciprocal of $\frac{3}{2}$ is ____
6. The reciprocal of 5 is ____
7. The reciprocal of $\frac{1}{4}$ is ____
8. The reciprocal of $\frac{5}{8}$ is ____
9. The reciprocal of $\frac{4}{9}$ is ____
10. The reciprocal of $\frac{7}{2}$ is ____
11. The reciprocal of $\frac{1}{9}$ is ____
12. The reciprocal of 10 is ____
13. The reciprocal of $\frac{7}{8}$ is ____
14. The reciprocal of $\frac{13}{15}$ is ____
15. The reciprocal of $\frac{5}{9}$ is ____
16. The reciprocal of $\frac{1}{6}$ is ____

Dividing Fractions

Divide each whole number by a fraction by multiplying by the reciprocal. If your answer is an improper fraction, change it to a whole or mixed number in lowest terms.

EX 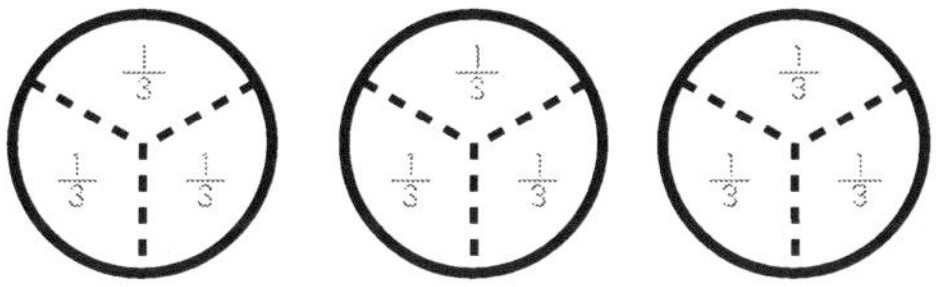

$3 \div \frac{1}{3} = 3 \times$ 3 = 9

1 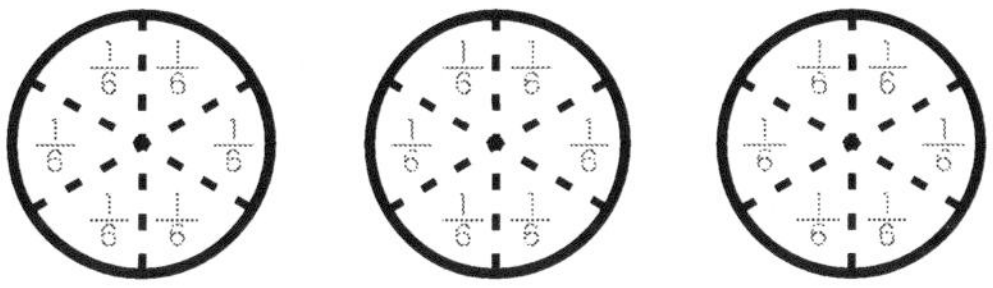

$3 \div \frac{1}{6} = 3 \times$ ____ = ______

2 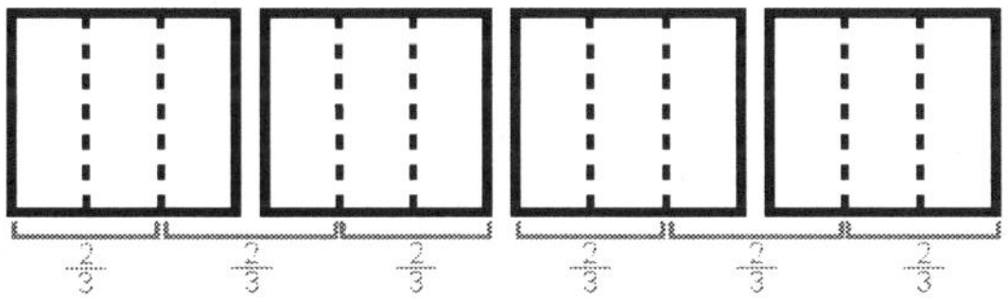

$4 \div \frac{2}{3} = 4 \times$ ____ = ______

3 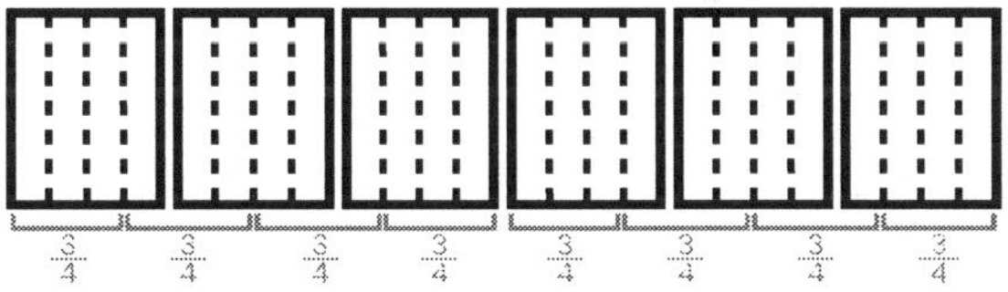

$6 \div \frac{3}{4} = 6 \times$ ____ = ______

4 $5 \div \frac{1}{4} = 5 \times$ ____ = ______

5 $2 \div \frac{1}{5} = 2 \times$ ____ = ______

6 $4 \div \frac{3}{8} = 4 \times$ ____ = ______

7 $9 \div \frac{3}{4} = 9 \times$ ____ = ______

8 $8 \div \frac{1}{6} = 8 \times$ ____ = ______

9 $5 \div \frac{5}{9} = 5 \times$ ____ = ______

10 $7 \div \frac{1}{3} =$ ______________

11 $3 \div \frac{1}{2} =$ ______________

12 $6 \div \frac{2}{3} =$ ______________

13 $10 \div \frac{2}{5} =$ ______________

84 Dividing Fractions

Divide each fraction by a whole number by multiplying by the reciprocal.
Write your answer in lowest terms.

EX 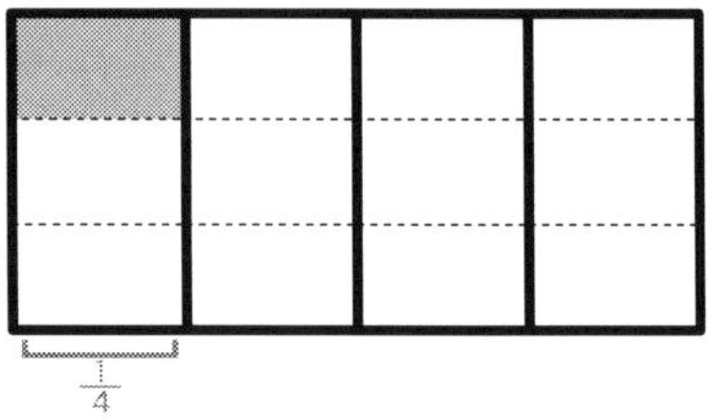

$\frac{1}{4} \div 3 = \frac{1}{4} \times \frac{1}{3} = \frac{1}{12}$

1 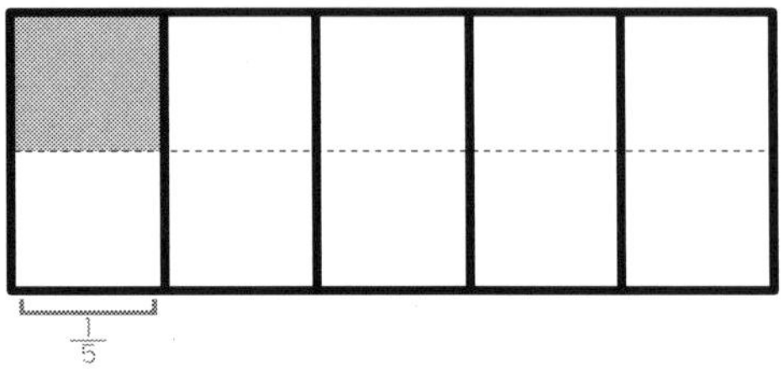

$\frac{1}{5} \div 2 = \frac{1}{5} \times$ ____ = ______

2 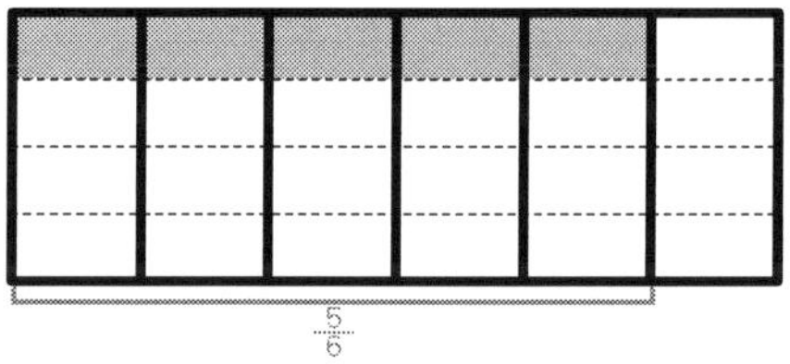

$\frac{5}{6} \div 4 = \frac{5}{6} \times$ ____ = ______

3 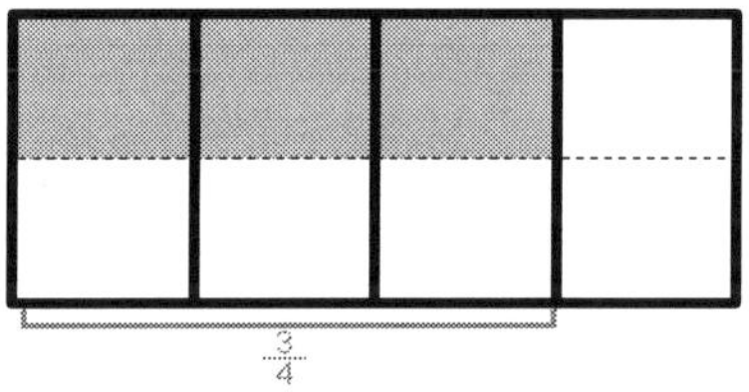

$\frac{3}{4} \div 2 = \frac{3}{4} \times$ ____ = ______

4 $\frac{1}{2} \div 6 = \frac{1}{2} \times$ ____ = ______

5 $\frac{1}{6} \div 5 = \frac{1}{6} \times$ ____ = ______

6 $\frac{3}{5} \div 2 = \frac{3}{5} \times$ ____ = ______

7 $\frac{2}{7} \div 4 = \frac{2}{7} \times$ ____ = ______

8 $\frac{3}{8} \div 6 = \frac{3}{8} \times$ ____ = ______

9 $\frac{4}{9} \div 3 = \frac{4}{9} \times$ ____ = ______

10 $\frac{5}{7} \div 2 =$ __________

11 $\frac{2}{3} \div 5 =$ __________

12 $\frac{4}{5} \div 8 =$ __________

13 $\frac{1}{3} \div 9 =$ __________

Dividing Fractions

Divide each pair of fractions by multiplying by the reciprocal of the second fraction. Express your answer as a fraction or mixed number in lowest terms.

(EX) $\frac{1}{2} \div \frac{2}{3} = \frac{1}{2} \times \frac{3}{2} = \frac{3}{4}$

(1) $\frac{3}{4} \div \frac{5}{6} = \frac{3}{4} \times$ ____ = ________

(2) $\frac{1}{4} \div \frac{3}{8} = \frac{1}{4} \times$ ____ = ________

(3) $\frac{1}{3} \div \frac{2}{5} = \frac{1}{3} \times$ ____ = ________

(4) $\frac{3}{5} \div \frac{1}{2} = \frac{3}{5} \times$ ____ = ________

(5) $\frac{4}{5} \div \frac{2}{3} = \frac{4}{5} \times$ ____ = ________

(6) $\frac{1}{6} \div \frac{1}{4} = \frac{1}{6} \times$ ____ = ________

(7) $\frac{5}{6} \div \frac{1}{3} = \frac{5}{6} \times$ ____ = ________

(8) $\frac{2}{7} \div \frac{3}{4} =$ ________________

(9) $\frac{4}{9} \div \frac{2}{5} =$ ________________

(10) $\frac{5}{8} \div \frac{1}{6} =$ ________________

(11) $\frac{2}{3} \div \frac{8}{9} =$ ________________

(12) $\frac{1}{8} \div \frac{2}{5} =$ ________________

(13) $\frac{3}{7} \div \frac{3}{4} =$ ________________

(14) $\frac{5}{9} \div \frac{1}{2} =$ ________________

(15) $\frac{3}{8} \div \frac{7}{9} =$ ________________

86 Dividing Fractions

Divide each pair of mixed numbers. Express your answer as a fraction or mixed number in lowest terms.

EX $2\frac{3}{5} \div 1\frac{1}{2} = \underline{1\frac{11}{15}}$

1. Before dividing, change any mixed numbers to improper fractions.

$2\frac{3}{5} = \frac{13}{5}$ $1\frac{1}{2} = \frac{3}{2}$

2. Divide by multiplying the first fraction by the reciprocal of the second fraction.

$\frac{13}{5} \div \frac{3}{2} = \frac{13}{5} \times \frac{2}{3} = \frac{26}{15}$

3. If needed, change the answer into a mixed number and simplify.

$\frac{26}{15} = 1\frac{11}{15}$

1. $4\frac{1}{3} \div 2\frac{5}{8} =$ ______

2. $3\frac{5}{9} \div 1\frac{2}{3} =$ ______

3. $3\frac{1}{2} \div 2\frac{3}{4} =$ ______

4. $4\frac{2}{5} \div 1\frac{4}{9} =$ ______

5. $2\frac{5}{12} \div 2\frac{1}{4} =$ ______

6. $5\frac{1}{3} \div 3\frac{3}{5} =$ ______

Dividing Fractions

Divide. Express your answer as a fraction or mixed number in lowest terms.

1. $4 \div \frac{3}{8} =$ ______
2. $\frac{5}{12} \div 3 =$ ______
3. $\frac{4}{7} \div \frac{5}{6} =$ ______
4. $4\frac{2}{5} \div 1\frac{4}{5} =$ ______
5. $6 \div 2\frac{1}{4} =$ ______
6. $\frac{7}{9} \div 3 =$ ______
7. $5 \div \frac{7}{8} =$ ______
8. $\frac{11}{12} \div \frac{2}{5} =$ ______
9. $4\frac{7}{8} \div 1\frac{2}{3} =$ ______
10. $5 \div 3\frac{1}{2} =$ ______

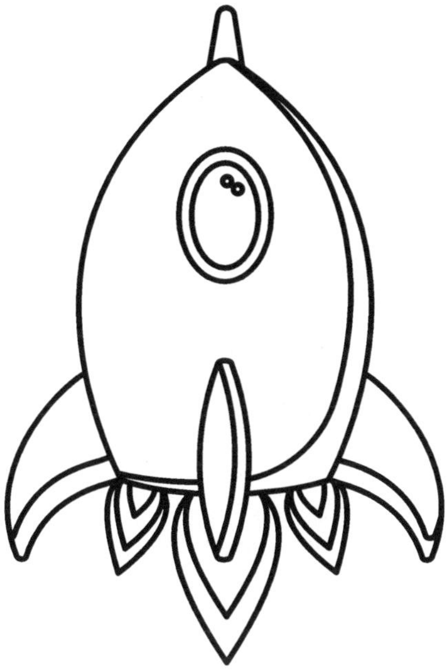

Decimals

90 Decimal Place Value

For each number, write the digits in the correct columns in the place value chart. Some columns may be empty for some numbers.

① 31.476

Hundreds	Tens	Ones	Tenths	Hundredths	Thousandths

② 103.6

Hundreds	Tens	Ones	Tenths	Hundredths	Thousandths

③ 5.79

Hundreds	Tens	Ones	Tenths	Hundredths	Thousandths

Write the value of the underlined digit in each number.

EX 6.1**9**1 ____0.09____

④ 43.**7**46 ______________

⑤ **8**.473 ______________

⑥ 19.25**4** ______________

⑦ 0.30**5** ______________

⑧ 6**7**8.53 ______________

⑨ 489.6**1**8 ______________

⑩ 5.**6**39 ______________

Fill in the blanks with the correct numbers.

⑪ In 248.563, the digit ______ is in the tenths place.

⑫ In 821.047 the digit ______ is in the thousandths place.

⑬ In 318.06, the digit ______ is in the tens place.

⑭ In 839.751, the digit ______ is in the hundredths place.

Decimal Place Value

Write each number in standard form.

EX Fifty-seven and sixty-three hundredths 57.63

1. One hundred two and four tenths ______
2. Four and nine hundred thirty-one thousandths ______
3. Sixteen and eight hundredths ______
4. Six hundred twenty-three thousandths ______

Write the name of each number.

EX 2.57 Two and fifty-seven hundredths

5. 15.235 ______
6. 43.79 ______
7. 17.03 ______
8. 723.5 ______

Write each number in expanded form.

EX 32.68 30 + 2 + 0.6 + 0.08

9. 426.7 ______
10. 0.978 ______
11. 5.264 ______

92 Decimal Place Value

Circle the number in each pair in which the digit **7** has a **greater** value.

1	32.786	34.097	2	0.576	5.76
3	47.153	70.98	4	52.764	47.328
5	6.075	8.007	6	174.5	2.765

Compare each pair of decimals using > or <.

EX	0.03	<	0.2	7	0.49	◯	0.325	8	1.67	◯	1.9
9	27.3	◯	2.73	10	9.03	◯	9.3	11	14.9	◯	15.8
12	7.01	◯	7.001	13	52.7	◯	52.08	14	9.35	◯	9.6

Rewrite each set of numbers in order from least to greatest.

15. 1.23 12.3 0.123 123 ______, ______, ______, ______

16. 0.48 0.048 0.84 0.084 ______, ______, ______, ______

17. 5.06 5.6 5.66 5.606 ______, ______, ______, ______

18. 7.654 7.456 7.564 7.645 ______, ______, ______, ______

Decimal Place Value

Fill in the chart to round each number to the given place value.

314.794	
Hundred	300
Ten	310
One	315
Tenth	314.8
Hundredth	314.79

1,249.816	
Hundred	
Ten	
One	
Tenth	
Hundredth	

853.179	
Hundred	
Ten	
One	
Tenth	
Hundredth	

94.352	
Hundred	
Ten	
One	
Tenth	
Hundredth	

6,285.143	
Hundred	
Ten	
One	
Tenth	
Hundredth	

712.528	
Hundred	
Ten	
One	
Tenth	
Hundredth	

94 Convert Decimals and Fractions

Rewrite each decimal as a fraction in lowest terms.

EX $0.26 = \frac{26}{100} = \frac{13}{50}$

1. 0.8 = ______________

2. 0.325 = ______________

3. 0.15 = ______________

4. 0.7 = ______________

5. 0.05 = ______________

6. 0.846 = ______________

7. 0.75 = ______________

8. 0.004 = ______________

9. 0.6 = ______________

10. 0.27 = ______________

11. 0.48 = ______________

---

12. 0.55 of the students in a school are boys. What fraction of the students are boys (in lowest terms)?

____

13. Gina walked 0.4 miles to her friend's house. What fraction of a mile did she walk?

____

Convert Decimals and Fractions

Rewrite each decimal as a mixed number in lowest terms.

EX $3.25 = 3\frac{25}{100} = 3\frac{1}{4}$

1. $14.5 =$ ____________

2. $2.125 =$ ____________

3. $7.96 =$ ____________

4. $10.3 =$ ____________

5. $6.08 =$ ____________

6. $1.508 =$ ____________

7. $21.2 =$ ____________

8. $5.005 =$ ____________

9. $9.36 =$ ____________

10. $3.85 =$ ____________

11. $201.7 =$ ____________

---

12. Kyle weighs 81.75 pounds. What is his weight expressed as a mixed number in lowest terms?

____

13. It takes the Moon 27.322 days to orbit Earth. What is the number of days expressed as a mixed number in lowest terms?

____

96 Convert Fractions and Decimals

Rewrite each fraction as a decimal.

EX $\frac{637}{1,000}$ = 0.637

1. $\frac{89}{100}$ = ______

2. $\frac{9}{10}$ = ______

3. $\frac{21}{100}$ = ______

4. $\frac{153}{1,000}$ = ______

5. $\frac{1}{10}$ = ______

6. $\frac{7}{100}$ = ______

7. $\frac{19}{1,000}$ = ______

8. $\frac{41}{100}$ = ______

9. $\frac{3}{1,000}$ = ______

EX $48\frac{73}{100}$ = 48.73

10. $16\frac{91}{100}$ = ______

11. $8\frac{397}{1,000}$ = ______

12. $81\frac{7}{10}$ = ______

13. $59\frac{1}{100}$ = ______

14. $141\frac{37}{100}$ = ______

15. $3\frac{71}{1,000}$ = ______

16. $23\frac{9}{1,000}$ = ______

Convert Fractions and Decimals

Change each fraction to an equivalent fraction with a denominator of 10, 100, or 1,000. Then, rewrite the fraction as a decimal.

EX $\frac{19}{20}$ = $\frac{95}{100}$ = 0.95

1. $\frac{1}{2}$ = ____ = ______

2. $\frac{3}{4}$ = ____ = ______

3. $\frac{5}{8}$ = ____ = ______

4. $\frac{2}{5}$ = ____ = ______

5. $\frac{1}{4}$ = ____ = ______

6. $\frac{43}{50}$ = ____ = ______

7. $\frac{239}{500}$ = ____ = ______

Change each mixed number to one with an equivalent fraction that has a denominator of 10, 100, or 1,000. Then, rewrite the mixed number as a decimal.

EX $8\frac{14}{25}$ = $8\frac{56}{100}$ = 8.56

8. $2\frac{11}{20}$ = ______ = ______

9. $5\frac{4}{5}$ = ______ = ______

10. $1\frac{231}{250}$ = ______ = ______

11. $9\frac{7}{25}$ = ______ = ______

12. $3\frac{33}{40}$ = ______ = ______

13. $7\frac{41}{50}$ = ______ = ______

14. $4\frac{2}{125}$ = ______ = ______

98 Add and Subtract Decimals

Find each sum.

EX $\begin{array}{r} 5.092 \\ +2.55 \\ \hline 7.642 \end{array}$

1. $\begin{array}{r} 7.23 \\ +1.465 \\ \hline \end{array}$

2. $\begin{array}{r} 6.182 \\ +0.706 \\ \hline \end{array}$

3. $\begin{array}{r} 4.331 \\ +2.9 \\ \hline \end{array}$

4. $\begin{array}{r} 0.861 \\ +0.073 \\ \hline \end{array}$

5. $\begin{array}{r} 5.1 \\ +6.325 \\ \hline \end{array}$

6. $\begin{array}{r} 2.946 \\ +3.15 \\ \hline \end{array}$

7. $\begin{array}{r} 3.708 \\ +4.21 \\ \hline \end{array}$

8. $\begin{array}{r} 12.37 \\ +\ \ 4.27 \\ \hline \end{array}$

9. $\begin{array}{r} 8.415 \\ +1.507 \\ \hline \end{array}$

10. $\begin{array}{r} 36.02 \\ +11.93 \\ \hline \end{array}$

11. $\begin{array}{r} 4.78 \\ +3.569 \\ \hline \end{array}$

12. 18.7 + 2.45 = ________

13. 5.08 + 19.736 = ________

14. 6.013 + 3.87 = ________

15. 10.98 + 2.15 = ________

16. 14.81 + 0.96 = ________

17. 102.17 + 37.2 = ________

Add and Subtract Decimals

Find each difference.

EX) $\begin{array}{r} 7.831 \\ -\ 5.22 \\ \hline 2.611 \end{array}$

1) $\begin{array}{r} 9.735 \\ -\ 6.504 \\ \hline \end{array}$

2) $\begin{array}{r} 8.625 \\ -\ 7.43 \\ \hline \end{array}$

3) $\begin{array}{r} 6.38 \\ -\ 0.151 \\ \hline \end{array}$

4) $\begin{array}{r} 5.028 \\ -\ 0.316 \\ \hline \end{array}$

5) $\begin{array}{r} 4.708 \\ -\ 2.56 \\ \hline \end{array}$

6) $\begin{array}{r} 7.15 \\ -\ 1.422 \\ \hline \end{array}$

7) $\begin{array}{r} 3.829 \\ -\ 0.235 \\ \hline \end{array}$

8) $\begin{array}{r} 40.54 \\ -\ 9.2 \\ \hline \end{array}$

9) $\begin{array}{r} 8.415 \\ -\ 1.507 \\ \hline \end{array}$

10) $\begin{array}{r} 23.91 \\ -\ 10.87 \\ \hline \end{array}$

11) $\begin{array}{r} 9.28 \\ -\ 4.063 \\ \hline \end{array}$

12) 87.4 - 24.9 = ________

13) 16.37 - 2.61 = ________

14) 5.12 - 4.8 = ________

15) 7.048 - 5.312 = ________

16) 25.64 - 3.241 = ________

17) 1.498 - 0.24 = ________

100 Add and Subtract Decimals

Fill in the missing numbers in each equation.

EX $\begin{array}{r} 3.1\boxed{3} \\ +6.42 \\ \hline 9.55 \end{array}$

1. $\begin{array}{r} 2.78 \\ +5.\square 1 \\ \hline 8.29 \end{array}$

2. $\begin{array}{r} 4.25 \\ +1.29 \\ \hline 5.\square 4 \end{array}$

3. $\begin{array}{r} 0.\square 8 \\ +3.47 \\ \hline 3.95 \end{array}$

4. $\begin{array}{r} 9.47 \\ -6.8\square \\ \hline 2.61 \end{array}$

5. $\begin{array}{r} 8.29 \\ -3.43 \\ \hline \square.86 \end{array}$

6. $\begin{array}{r} \square.05 \\ -1.24 \\ \hline 5.81 \end{array}$

7. $\begin{array}{r} 5.76 \\ -\square.12 \\ \hline 2.64 \end{array}$

8. $\begin{array}{r} 1.08 \\ +4.98 \\ \hline \square.06 \end{array}$

9. $\begin{array}{r} 4.6\square \\ -0.49 \\ \hline 4.13 \end{array}$

10. $\begin{array}{r} 2.73 \\ +5.\square 1 \\ \hline 8.34 \end{array}$

11. $\begin{array}{r} 6.43 \\ -2.75 \\ \hline 3.6\square \end{array}$

Compare each pair of expressions using >, <, or =.

12. $0.03 + 2.751 \bigcirc 1.37 + 1.506$

13. $4.801 - 1.52 \bigcirc 5.6 - 2.982$

14. $6.23 + 0.89 \bigcirc 9.85 - 2.36$

15. $7.513 - 4.08 \bigcirc 1.48 + 1.953$

Add and Subtract Decimals

Add or subtract to solve each word problem.

1. After work, Jason drives 3.8 miles to the library, and then another 6.25 miles to get home. How far did he drive in all?

2. In a 50-yard dash, Emily finished in 6.37 seconds and Kayla finished in 5.9 seconds. By how much did Kayla beat Emily in the race?

3. Dominic buys a book that costs $11.95. He pays with a $20 bill. How much change should he get back?

4. A city received 3.02 inches of rain in March and 3.18 inches of rain in April. How much rain did the city receive across both months?

5. Kelsey orders a burger that costs $6.29 and a drink that costs $1.99. What is the total price of her meal?

102 Multiplying Decimals

Multiply each decimal by a power of ten. For each zero in the power of ten, move the decimal point one place value to the right.

(EX) $2.376 \times 100 =$ 237.6

100 has two zeroes, so move the decimal point two places to the right.

2376

(1) $1.25 \times 10 =$ ________

(2) $52.17 \times 100 =$ ________

(3) $4.26 \times 1{,}000 =$ ________

(4) $49.35 \times 10 =$ ________

(5) $0.289 \times 100 =$ ________

(6) $16.48 \times 10{,}000 =$ ________

Multiply each decimal by a power of ten. Move the decimal point the number of places to the right indicated by the exponent.

(EX) $0.264 \times 10^3 =$ 264

The exponent is 3, so move the decimal point three places to the right.

0264

(7) $8.264 \times 10^3 =$ ________

(8) $29.3 \times 10^2 =$ ________

(9) $1.687 \times 10^4 =$ ________

(10) $2.087 \times 10^2 =$ ________

(11) $64.781 \times 10^3 =$ ________

(12) $0.092 \times 10^4 =$ ________

Multiplying Decimals

Multiply each decimal by a whole number.

EX

$$\begin{array}{r} 1.23 \\ \times\ 45 \\ \hline 615 \\ +\ 4920 \\ \hline 55.35 \end{array}$$

1. Multiply as you would with two whole numbers.
2. Count up the decimals places in the factors.
3. Move the decimal point in the product left that many places.

① $\begin{array}{r} 21.3 \\ \times\ 8 \\ \hline \end{array}$

② $\begin{array}{r} 5.75 \\ \times\ 16 \\ \hline \end{array}$

③ $\begin{array}{r} 0.87 \\ \times\ 26 \\ \hline \end{array}$

④ $\begin{array}{r} 6.54 \\ \times\ 37 \\ \hline \end{array}$

⑤ $\begin{array}{r} 0.97 \\ \times\ 64 \\ \hline \end{array}$

⑥ $\begin{array}{r} 15.6 \\ \times\ 9 \\ \hline \end{array}$

⑦ $\begin{array}{r} 8.72 \\ \times\ 7 \\ \hline \end{array}$

⑧ $\begin{array}{r} 1.75 \\ \times\ 48 \\ \hline \end{array}$

⑨ $\begin{array}{r} 51.4 \\ \times\ 11 \\ \hline \end{array}$

Multiplying Decimals

Use the area model to help you multiply each pair of decimals.

(EX)

0.2

0.8

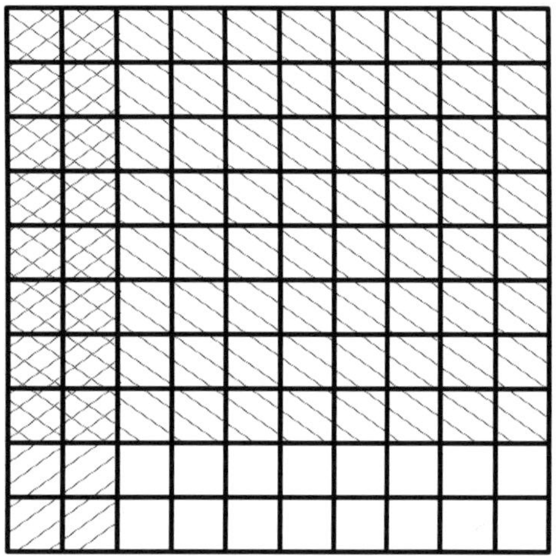

0.2 × 0.8 = 0.16

(1)

0.1

0.9

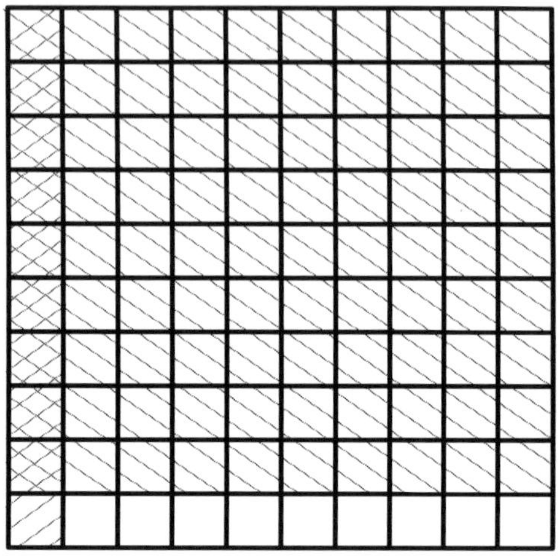

0.1 × 0.9 = ________

(2)

0.3

0.7

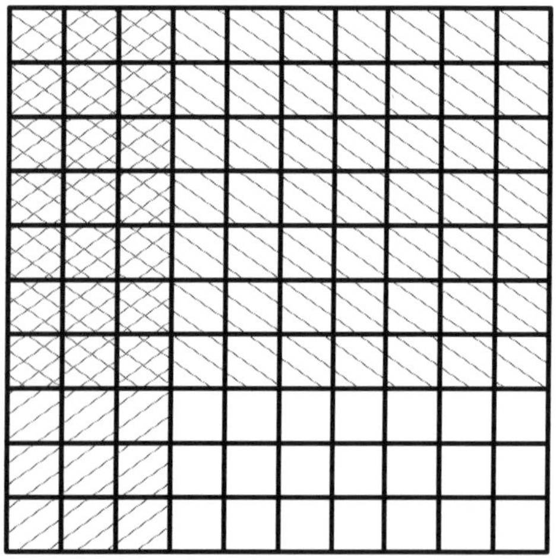

0.3 × 0.7 = ________

(3)

0.4

0.6

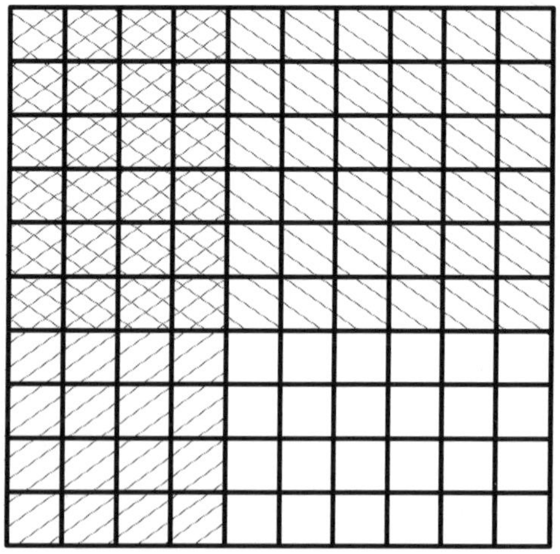

0.4 × 0.6 = ________

Multiplying Decimals

Shade in the area model and multiply each pair of decimals.

1

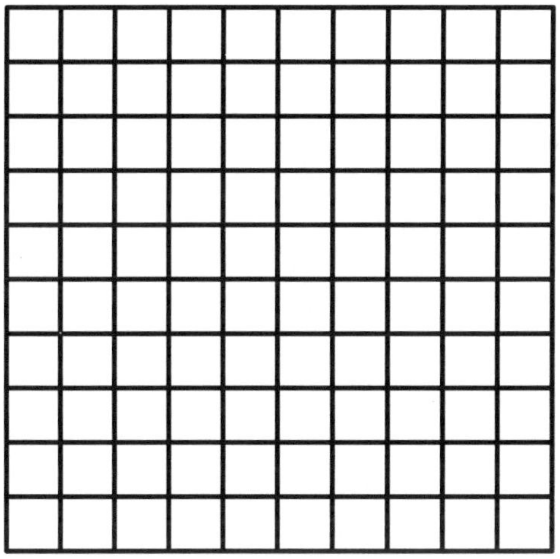

$0.5 \times 0.9 =$ ________

2

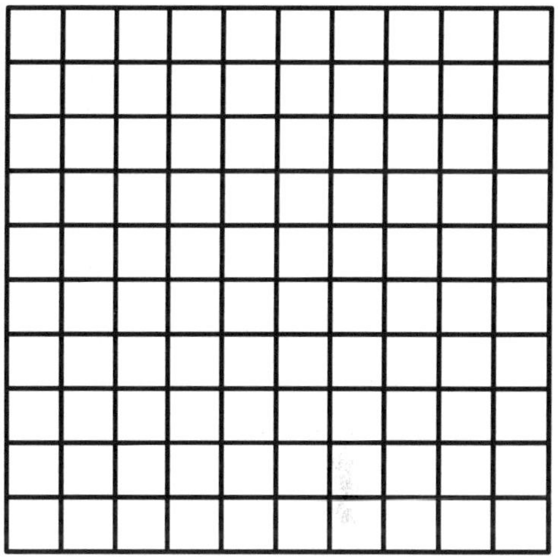

$0.8 \times 0.3 =$ ________

3

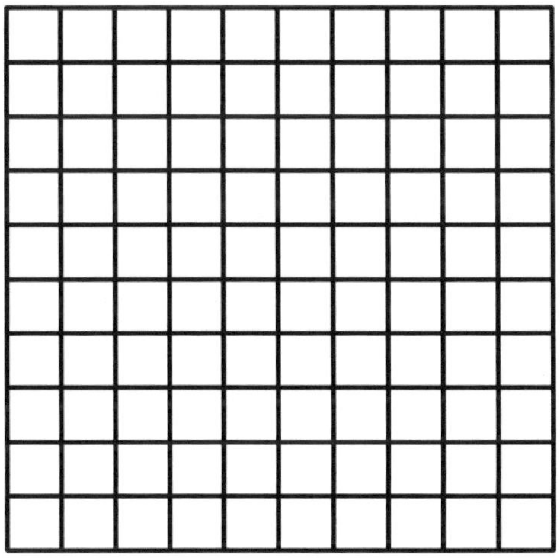

$0.7 \times 0.4 =$ ________

4

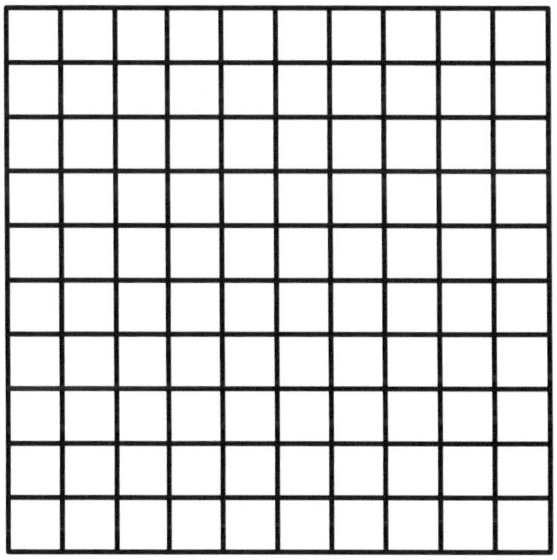

$0.6 \times 0.2 =$ ________

106 Multiplying Decimals

Multiply each pair of decimals.

EX

$$\begin{array}{r} \overset{2\ 1}{1.43} \\ \times\ 2.5 \\ \hline 715 \\ +\ 2860 \\ \hline 3.575 \end{array}$$

1. Multiply as you would with two whole numbers.
2. Count up the decimals places in both factors.
3. Move the decimal point in the product left that many places.

(1) $\begin{array}{r} 16.3 \\ \times\ 0.2 \\ \hline \end{array}$

(2) $\begin{array}{r} 4.72 \\ \times\ 3.6 \\ \hline \end{array}$

(3) $\begin{array}{r} 5.37 \\ \times\ 8.9 \\ \hline \end{array}$

(4) $\begin{array}{r} 10.8 \\ \times\ 1.26 \\ \hline \end{array}$

(5) $\begin{array}{r} 2.93 \\ \times\ 4.4 \\ \hline \end{array}$

(6) $\begin{array}{r} 95.13 \\ \times\ 0.6 \\ \hline \end{array}$

(7) $\begin{array}{r} 19.8 \\ \times\ 0.7 \\ \hline \end{array}$

(8) $\begin{array}{r} 9.7 \\ \times\ 6.8 \\ \hline \end{array}$

(9) $\begin{array}{r} 21.34 \\ \times\ 3.2 \\ \hline \end{array}$

Multiplying Decimals

Multiply.

1. $\begin{array}{r} 4.86 \\ \times\ 23 \\ \hline \end{array}$

2. $\begin{array}{r} 31.5 \\ \times\ 7.6 \\ \hline \end{array}$

3. $\begin{array}{r} 8.19 \\ \times\ 5.2 \\ \hline \end{array}$

4. $0.018 \times 10^3 =$ ______

5. $4.03 \times 100 =$ ______

6. $0.4 \times 0.2 =$ ______

7. $8 \times 0.06 =$ ______

Multiply to solve each word problem.

8. A brick is 2.25 inches tall. If 7 bricks were stacked on top of each other, what would the height of the stack be?

9. At the school store, a pen costs $0.85. How much would 5 pens cost?

10. A rug measures 3.5 meters long by 2.5 meters wide. What is the area of the rug, in square meters?

108 Dividing Decimals

Divide each decimal by a power of ten. For each zero in the power of ten, move the decimal point one place value to the left.

(EX) $324.9 \div 100 =$ 3.249

100 has two zeroes, so move the decimal point two places to the left.

324.9

1. $57.9 \div 10 =$ ________
2. $3.1 \div 100 =$ ________
3. $673.2 \div 1{,}000 =$ ________
4. $156.52 \div 10 =$ ________
5. $16.79 \div 100 =$ ________
6. $9{,}875.3 \div 10{,}000 =$ ________

Divide each decimal by a power of ten. Move the decimal point the number of places to the left indicated by the exponent.

(EX) $875.9 \div 10^3 =$ 0.8759

The exponent is 3, so move the decimal point three places to the left.

875.9

7. $1{,}985.2 \div 10^3 =$ ________
8. $5.76 \div 10^2 =$ ________
9. $975.34 \div 10^4 =$ ________
10. $12.823 \div 10^2 =$ ________
11. $3.1 \div 10^3 =$ ________
12. $0.9 \div 10^4 =$ ________

Dividing Decimals

Divide each decimal by a whole number.

EX

$$\begin{array}{r} 3.6 \\ 12\overline{)43.2} \\ -36\downarrow \\ \hline 72 \\ -72 \\ \hline 0 \end{array}$$

1. Put a decimal point in the quotient directly above the decimal point in the dividend.
2. Divide as you would normally with whole numbers.

1. $8\overline{)24.96}$
2. $7\overline{)183.4}$
3. $15\overline{)18.45}$
4. $3\overline{)172.2}$
5. $11\overline{)52.25}$
6. $4\overline{)92.8}$
7. $6\overline{)1.44}$
8. $9\overline{)65.7}$
9. $5\overline{)81.5}$

Dividing Decimals

Divide. Express any partial quotients as a decimal rather than a remainder.

EX $8\overline{)298.00}$ = 37.25

```
   37.25
8)298.00
 -24
   58
  -56
    20
   -16
     40
    -40
      0
```

If the numbers do not divide evenly, add a decimal point to the end of the dividend (if it does not already have one) and attach a zero. Continue to divide. Add additional zeroes and repeat as needed until there is no remainder.

EX $2\overline{)27.50}$ = 13.75

```
   13.75
2)27.50
 -2
  07
  -6
   15
  -14
    10
   -10
     0
```

2. $5\overline{)82}$

3. $4\overline{)302}$

4. $12\overline{)303}$

5. $6\overline{)14.7}$

6. $8\overline{)51.6}$

7. $2\overline{)67.7}$

Dividing Decimals

Divide each pair of decimals.

(EX) 5.75 ÷ 0.46 = 12.5

1. When the divisor is a decimal, change it to a whole number by multiplying by a power of 10. Multiply the divisor by the same power of 10.

0.46 × 100 = 46

5.75 × 100 = 575

2. Divide.

```
     12.5
46 )575.0
   -46
    115
    -92
     230
    -230
       0
```

(1) 74.88 ÷ 1.2 = ________

(2) 22.26 ÷ 0.7 = ________

(3) 2.96 ÷ 0.08 = ________

(4) 10.452 ÷ 0.26 = ________

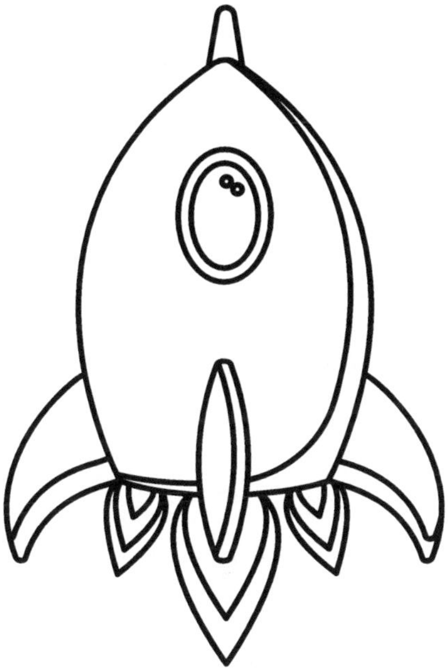

Measurement and Data

114 Customary Measurement Units

Convert between customary units of length.

(1) 1 yard = _______ inches

(2) 1 foot = _______ inches

(3) 1 mile = _______ feet

(4) 1,760 yards = _______ miles

(5) 12 feet = _______ yards

(6) 72 inches = _______ feet

(7) 3 yards = _______ inches

(8) $\frac{1}{2}$ mile = _______ feet

(9) 2 miles = _______ yards

(10) $\frac{2}{3}$ yard = _______ inches

(11) $2\frac{1}{2}$ feet = _______ inches

(12) $\frac{1}{4}$ yard = _______ inches

Compare each pair of lengths using >, <, or =.

(EX) 35 in. (<) 1 yd.

(13) 1 mi. ◯ 5,820 ft.

(14) 47 in. ◯ 4 ft.

(15) 16 in. ◯ $\frac{1}{2}$ yd.

(16) 26 ft. ◯ 8 yd.

(17) 10 ft. ◯ $3\frac{1}{3}$ yd.

Customary Measurement Units

Convert between customary units of weight.

(1) 1 pound = ________ ounces

(2) 1 ton = ________ pounds

(3) 32 ounces = ________ pounds

(4) 6,000 pounds = ______ tons

(5) 1 ton = ________ ounces

(6) 5 pounds = ________ ounces

(7) $\frac{1}{2}$ ton = ________ pounds

(8) $\frac{1}{4}$ pound = ________ ounces

(9) $\frac{3}{4}$ ton = ________ pounds

(10) 2 tons = ________ ounces

(11) $4\frac{1}{2}$ tons = ________ pounds

(12) $3\frac{1}{8}$ pounds = ______ ounces

Compare each pair of weights using >, <, or =.

(EX) 31 oz. (<) 2 lbs.

(13) 8 T ( ) 15,000 lbs.

(14) $\frac{7}{8}$ lb. ( ) 14 oz.

(15) $\frac{1}{2}$ T ( ) 16,000 oz.

(16) 12 lbs. ( ) 200 oz.

(17) 80 oz. ( ) $5\frac{1}{2}$ lbs.

116 Customary Measurement Units

Convert between customary units of volume.

(1) 1 pint = _______ cups

(2) 1 quart = _______ cups

(3) 1 cup = _______ fluid ounces

(4) 1 gallon = _______ quarts

(5) 1 quart = _______ pints

(6) 20 quarts = _______ gallons

(7) 2 quarts = _______ cups

(8) $\frac{1}{2}$ gallon = _______ cups

(9) 3 pints = _______ fluid ounces

(10) 72 pints = _______ gallons

(11) $6\frac{1}{2}$ quarts = _______ pints

(12) $\frac{1}{2}$ pint = _____ fluid ounces

Compare each pair of volumes using >, <, or =.

(EX) 10 pt. (<) 10 qt.

(13) 1 gal. ( ) 64 fl. oz.

(14) 20 c. ( ) 10 pt.

(15) $\frac{1}{4}$ qt. ( ) 6 fl. oz.

(16) 8 qt. ( ) 24 c.

(17) 9 pt. ( ) $1\frac{1}{8}$ gal.

Customary Measurement Units

Convert between mixed customary units of measurement.

EX 40 inches = 3 feet 4 inches

1. 66 ounces = _____ pounds _____ ounces

2. 45 fluid ounces = _____ cups _____ fluid ounces

3. 26 feet = _____ yards _____ feet

4. 7,356 pounds = _____ tons _____ pounds

5. 18 pints = _____ gallons _____ pints

6. 1 mile 87 feet = _______ feet

7. 8 pounds 10 ounces = _____ ounces

8. 7 quarts 2 cups = _____ cups

9. 6 feet 3 inches = _____ inches

10. 9 cups 6 fluid ounces = _____ fluid ounces

118 Customary Measurement Units

Add mixed customary units of measurement.

(EX) 3 qt. 1 pt. + 2 qt. 1 pt. = 6 qt. 0 pt.

1. Add like units.

	3 qt.	1 pt.
+	2 qt.	1 pt.
	5 qt.	2 pt.

2. Regroup if needed. 6 qt. 0 pt. Since 2 pt. = 1 qt., you can regroup.

(1) 8 lbs 3 oz. + 6 lbs. 9 oz. = _____ lbs. _____ oz.

(2) 4 c. 5 fl. oz. + 2 c. 6 fl. oz. = _____ c. _____ fl. oz.

(3) 5 yd. 2 ft. + 7 yd. 2 ft. = _____ yd. _____ ft.

(4) A delivery truck weighs 5 T when empty. When it is carrying a load that weighs 2 T 400 lbs., what is its total weight?

Subtract mixed customary units of measurement.

EX 2 mi. 300 ft. - 800 ft. = 1 mi. 4,780 ft.

1. Borrow if needed.

	2 mi.	300 ft.
-		800 ft.

1 mi. = 5,280 ft., so you can borrow 1 from the miles and add 5,280 to the feet.

2. Subtract like units.

	1 mi.	5,580 ft.
-		800 ft.
	1 mi.	4,780 ft.

1. 10 yd. 1 ft. - 5 yd. 2 ft. = _____ yd. _____ ft.

2. 2 T - 500 lbs. = _____ T _____ lbs.

3. 4 gal. - 15 c. = _____ gal. _____ c.

4. James is 6 ft. 2 in. tall. Morgan is 5 ft. 6 in. tall. How much taller is James than Morgan?

120 Metric Measurement Units

Use the metric unit conversion chart to help you convert between metric units of length.

Kilo-	Hecto-	Deka-	Base Unit	Deci-	Centi-	Milli-
1,000	100	10	1	0.1	0.01	0.001
Kilometer (km)	Hectometer (hm)	Dekameter (dam)	Meter (m)	Decimeter (dm)	Centimeter (cm)	Millimeter (mm)

1. 3 m = ________ cm
2. 7 km = ________ m
3. 60 dam = ________ hm
4. 12 dm = ________ mm
5. 350 mm = ________ cm
6. 27 hm = ________ km
7. 183 cm = ________ dam
8. 8.9 m = ________ mm
9. 16.5 hm = ________ dm
10. 0.2 dam = ________ km

Compare each pair of lengths using >, <, or =.

EX. 37 cm (<) 4 m

11. 8.5 hm ◯ 850 m
12. 0.3 km ◯ 30 dam
13. 91 dm ◯ 700 m
14. 14 mm ◯ 1 cm
15. 540 m ◯ 5.5 km

Metric Measurement Units

Use the metric unit conversion chart to help you convert between metric units of mass.

Kilo-	Hecto-	Deka-	Base Unit	Deci-	Centi-	Milli-
1,000	100	10	1	0.1	0.01	0.001
Kilogram (kg)	Hectogram (hg)	Dekagram (dag)	Gram (g)	Decigram (dg)	Centigram (cg)	Milligram (mg)

1. 14 g = ________ dag
2. 11 kg = ________ hg
3. 3.2 dag = ________ cg
4. 240 dg = ________ mg
5. 98.3 mg = ________ g
6. 13.1 hg = ________ dag
7. 0.4 cg = ________ dg
8. 27 g = ________ mg
9. 1.37 hg = ________ dg
10. 4 dag = ________ hg

Circle the largest mass in each row.

11	43 g	430 cg	0.43 kg	4,300 mg
12	8.5 dag	850 hg	0.085 kg	8,500 cg
13	239 dg	2.39 kg	0.239 hg	239 g
14	78.2 cg	782 g	7,820 mg	7.82 dg

122 Metric Measurement Units

Use the metric unit conversion chart to help you convert between metric units of volume.

Kilo-	Hecto-	Deka-	Base Unit	Deci-	Centi-	Milli-
1,000	100	10	1	0.1	0.01	0.001
Kiloliter (kL)	Hectoliter (hL)	Dekaliter (daL)	Liter (L)	Deciliter (dL)	Centiliter (cL)	Milliliter (mL)

1. 28 L = ________ mL
2. 1.7 kL = ________ L
3. 41 daL = ________ hL
4. 0.9 dL = ________ mL
5. 721 mL = ________ cL
6. 3.7 hL = ________ kL
7. 92.1 cL = ________ daL
8. 22 L = ________ cL
9. 63 hL = ________ dL
10. 70 daL = ________ kL

Circle the smallest volume in each row.

11	87 L	870 mL	87,000 cL	8.7 daL
12	5.23 daL	52.3 hL	5.23 kL	523 L
13	9.12 dL	0.912 daL	91.2 L	912 cL
14	0.05 hL	0.5 L	5 cL	500 mL

Metric Measurement Units

Convert between mixed metric units of measurement.

EX 4 m 50 cm = 450 cm

1. 8 L 25 mL = ______ mL

2. 13 kg 450 g = ______ g

3. 2,134 m = _____ km _____ m

4. 12,936 mL = _____ L ______ mL

5. 6,503 mg = _____ g _____ mg

6. 10 hm 87 m = ______ m

7. 3 L 8 dL = ______ dL

8. 31 g 45 cg = ______ cg

9. 15,927 mm = ______ m ______ mm

10. 502 L = ______ hL ______ L

124 Metric Measurement Units

Add mixed metric units of measurement.

(EX) 3 kg 358 g + 2 kg 722 g = 6 kg 80 g

1. Add like units.

$$\begin{array}{rr} 3\text{ kg} & 358\text{ g} \\ +\ 2\text{ kg} & 722\text{ g} \\ \hline 5\text{ kg} & 1{,}080\text{ g} \end{array}$$

2. Regroup if needed. 6 kg 80 g

Since 1,000 g = 1 kg, you can regroup. Subtract 1,000 from the g and add 1 to the kg.

(1) 8 m 47 cm + 4 m 58 cm = _____ m _____ cm

(2) 9 L 201 mL + 5 L 63 mL = _____ L _____ mL

(3) 3 g 27 mg + 5 g 102 mg = _____ g _____ mg

(4) Allison made punch that included 3 L 700 mL of juice and 1 L 500 mL of ginger ale. What is the total volume of liquid in the punch?

Subtract mixed metric units of measurement.

(EX) 7 L 200 mL - 3 L 650 mL = 3 L 550 mL

1. Borrow if needed.

7 L	200 mL
- 3 L	650 mL

1 L = 1,000 mL, so you can borrow 1 from the liters and add 1,000 to the milliliters.

2. Subtract like units.

6 L	1,200 mL
- 3 L	650 mL
3 L	550 mL

(1) 4 km 300 m - 2 km 400 m = _____ km _____ m

(2) 6 kg - 700 g = _____ kg _____ g

(3) 5 L 250 mL - 1 L 186 mL = _____ L _____ mL

(4) Kyle ran 4 km 500 m on Tuesday. On Wednesday, he ran 3 km 800 m. How much farther did he run on Tuesday than Wednesday?

Find the perimeter of each figure.

1 Perimeter = ______ units

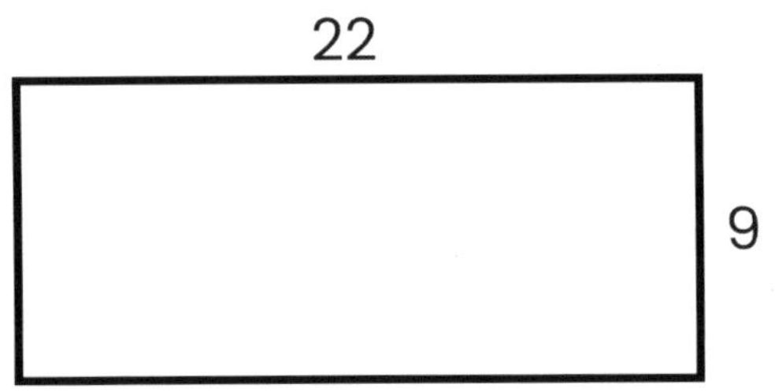

2 Perimeter = ______ units

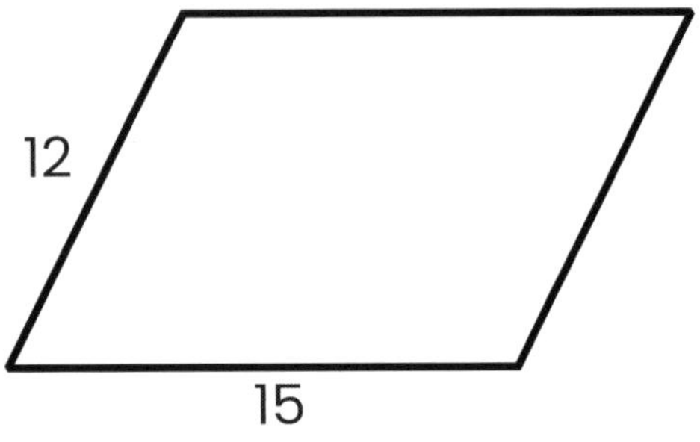

3 Perimeter = ______ units

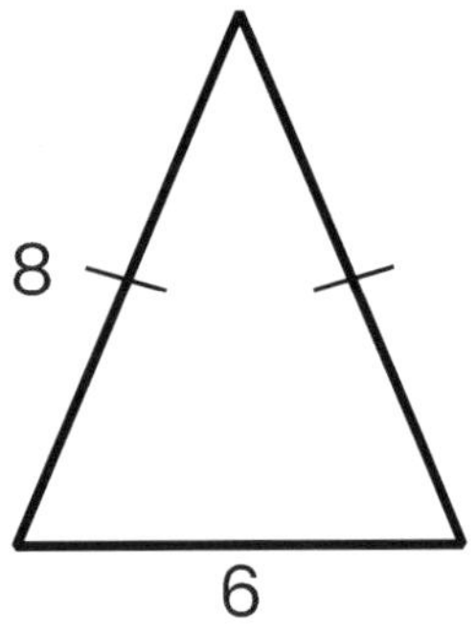

4 Perimeter = ______ units

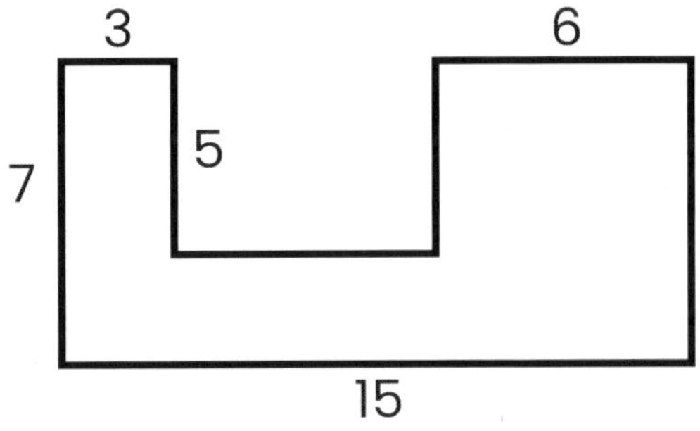

5 Perimeter = ______ units

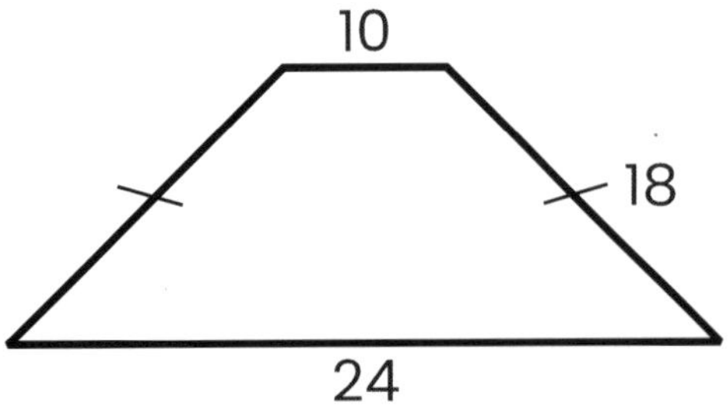

6 Perimeter = ______ units

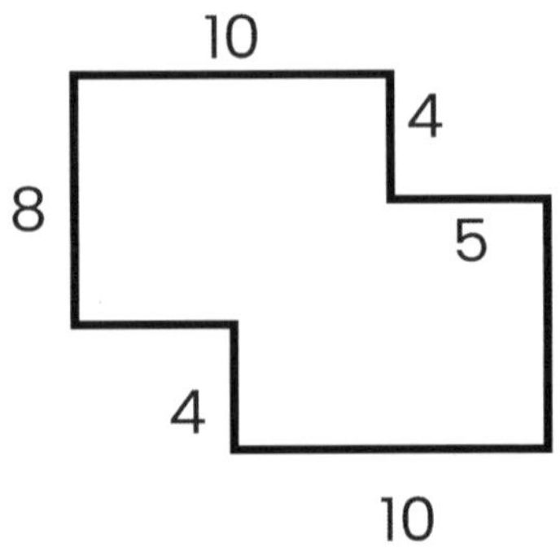

Find the perimeter. Write your answer as a whole or mixed number in lowest terms.

1. Perimeter = ______ in.

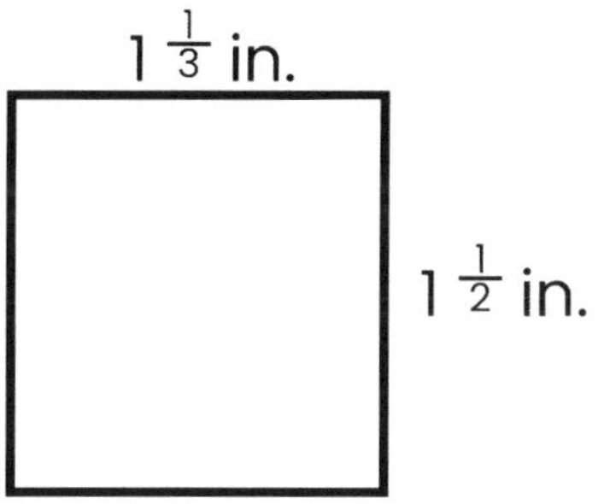

2. Perimeter = ______ m

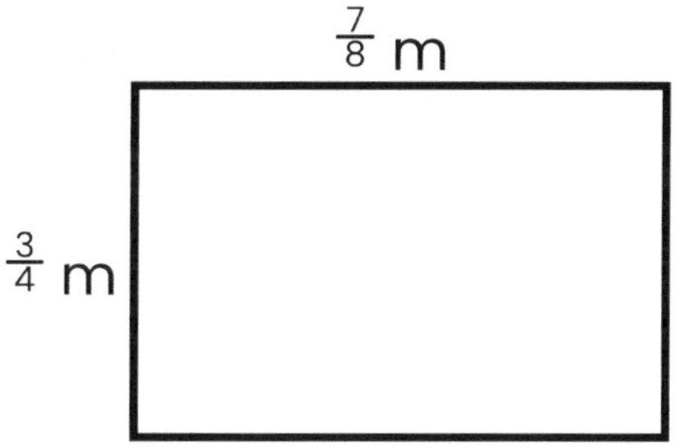

3. Perimeter = ______ cm

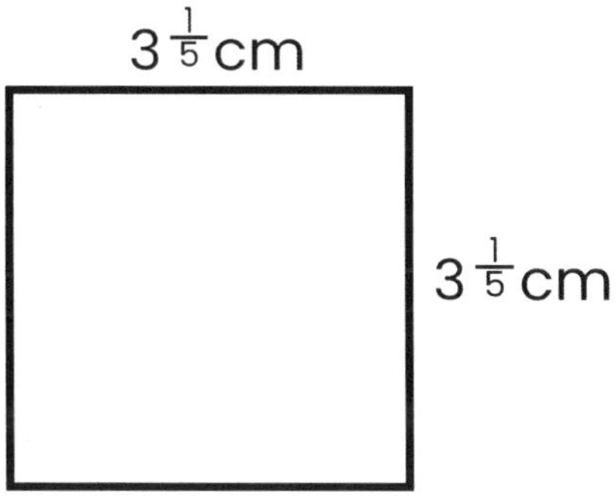

4. Perimeter = ______ ft.

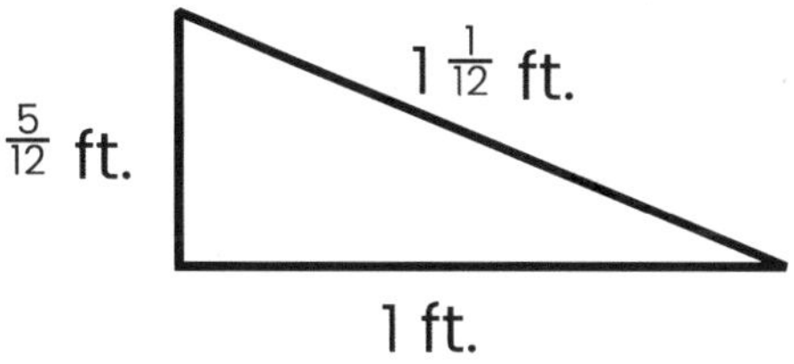

5. A piece of paper measures $8\frac{1}{2}$ inches by 11 inches. What is the perimeter of the paper?

128 Geometric Measurement

Find the perimeter. Write your answer as a whole number or decimal.

(1) Perimeter = ______ in.

(2) Perimeter = ______ m

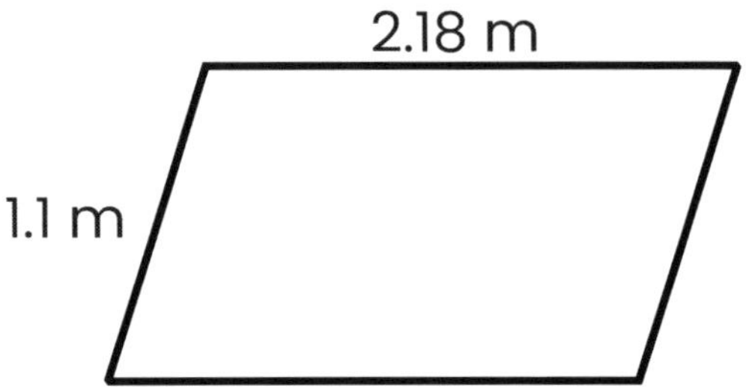

(3) Perimeter = ______ cm

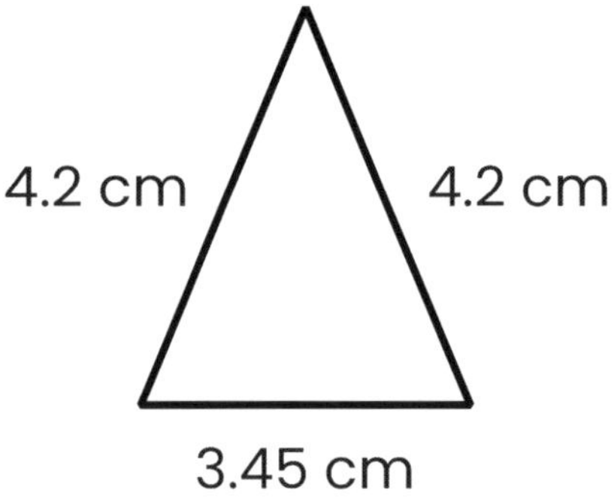

(4) Perimeter = ______ ft.

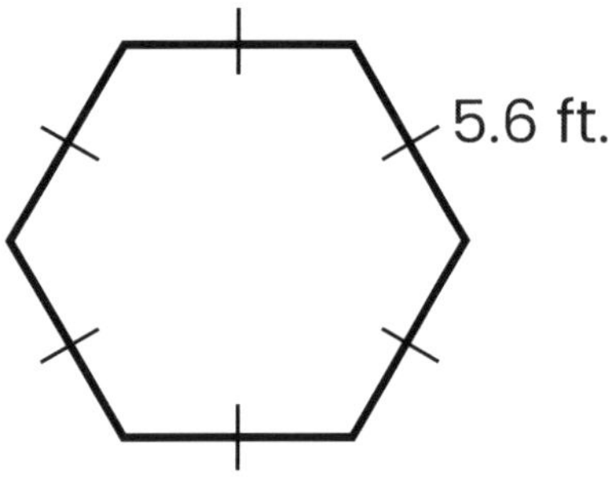

(5) A square garden plot measures 3.25 meters on each side. What is the perimeter of the plot?

Use the perimeter of each figure to find the length of the side marked x.

① Perimeter = 50 units

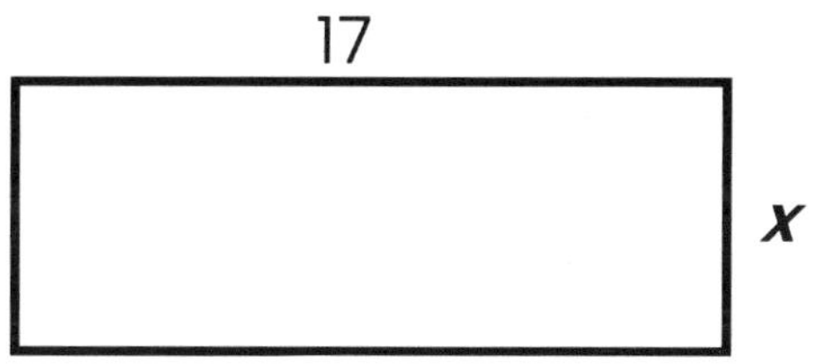

x = ______ units

② Perimeter = 35 units

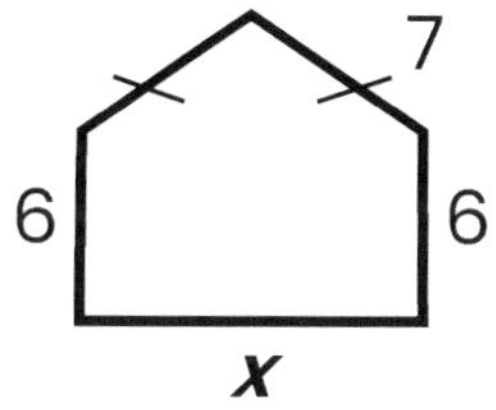

x = ______ units

③ Perimeter = 36 units

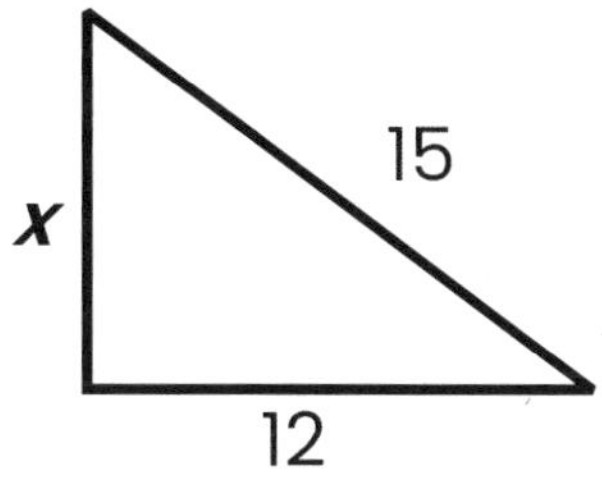

x = ______ units

④ Perimeter = 80 units

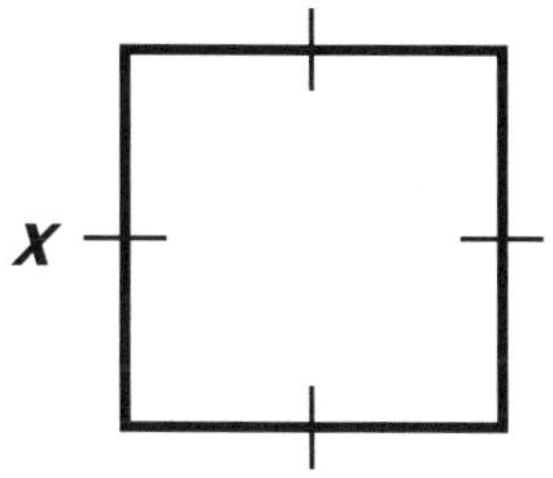

x = ______ units

⑤ Perimeter = 65 units

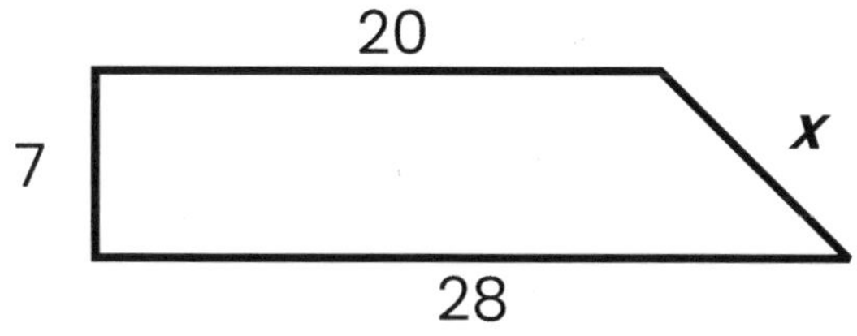

x = ______ units

⑥ Perimeter = 45 units

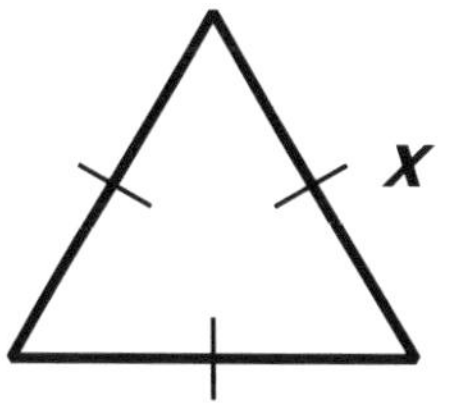

x = ______ units

130 Geometric Measurement

Find the area of each rectangle.

1. Area = ________ cm^2

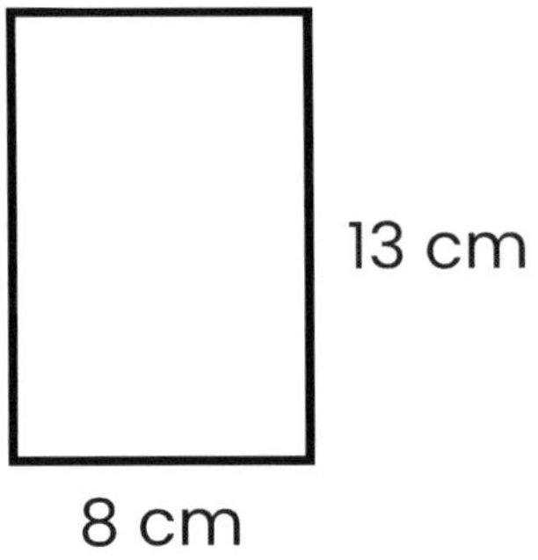

2. Area = ________ ft^2

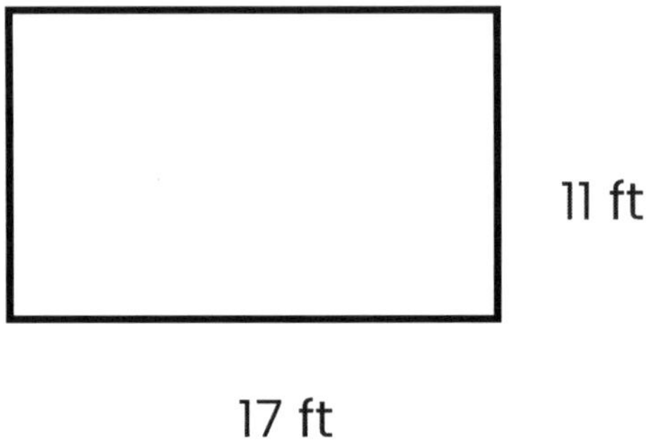

3. Area = ________ in^2

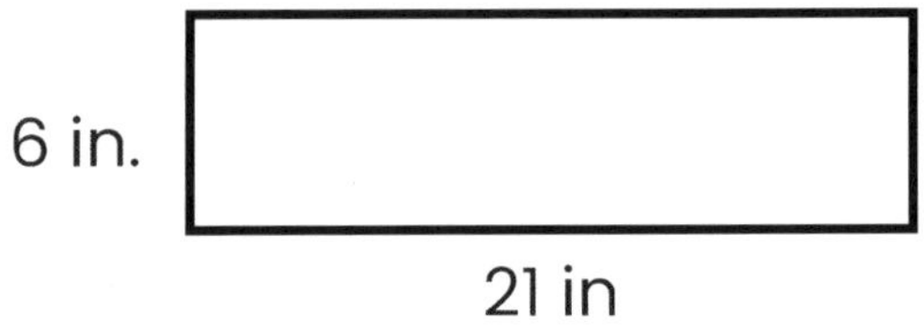

4. Area = ________ m^2

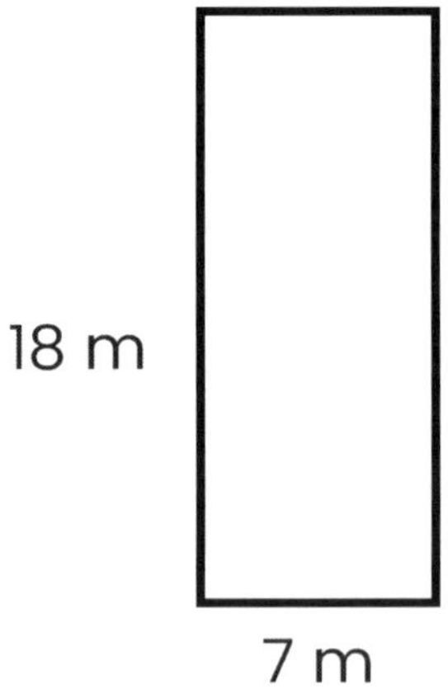

5. A floor tile measures 12 inches by 24 inches. What is the area of the tile in square inches?

Find the area of each rectangle. Write your answer as a whole or mixed number in lowest terms.

1. Area = ________ in²

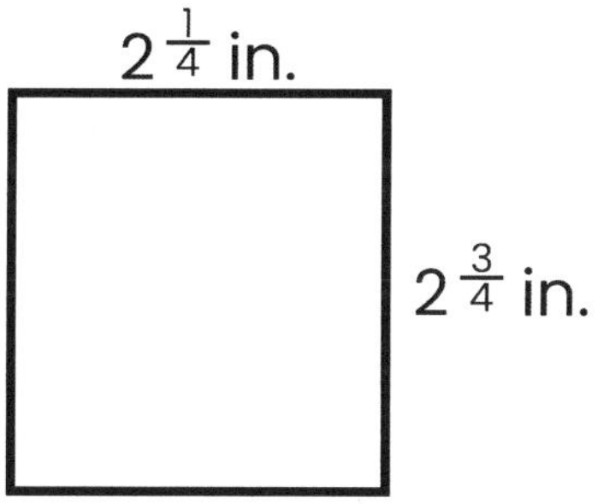

2. Area = ________ ft²

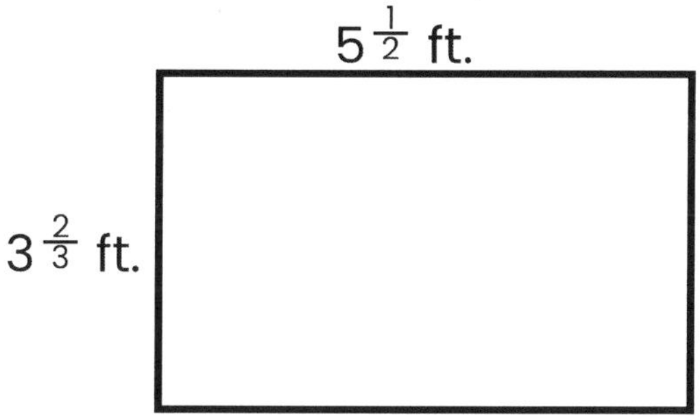

3. Area = ________ cm²

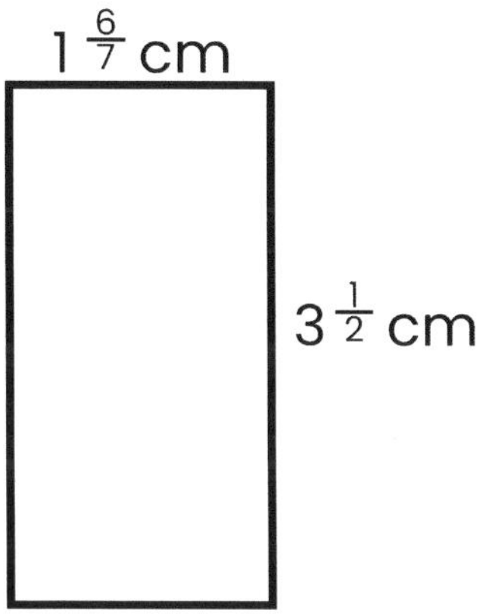

4. Area = ________ m²

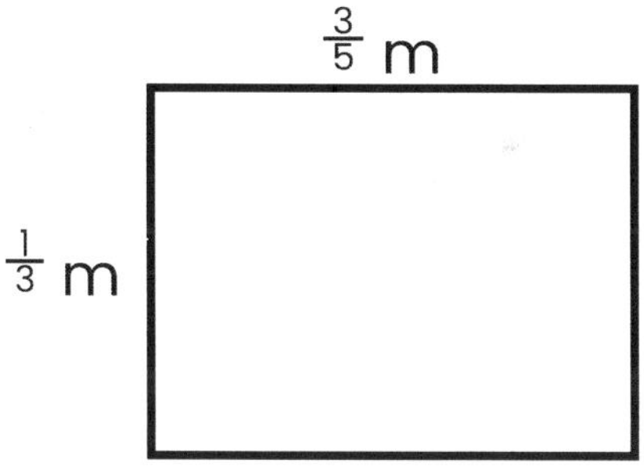

5. A rectangular rug measures $6\frac{1}{2}$ feet by $9\frac{1}{4}$ feet. What is the area of the rug in square feet?

Geometric Measurement

Find the area of each rectangle. Write your answer as a whole number or decimal.

1. Area = ________ cm^2

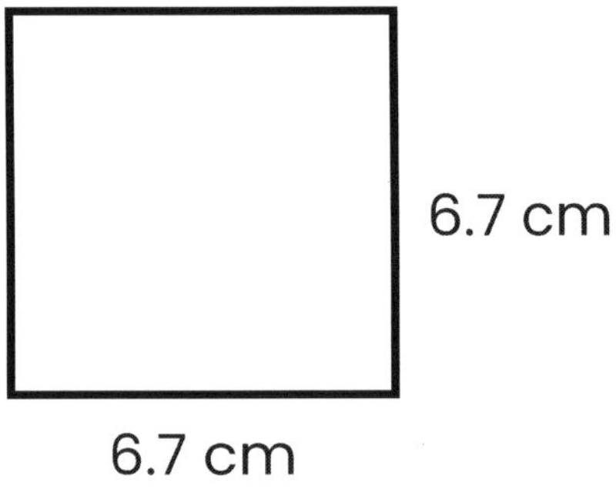

2. Area = ________ ft^2

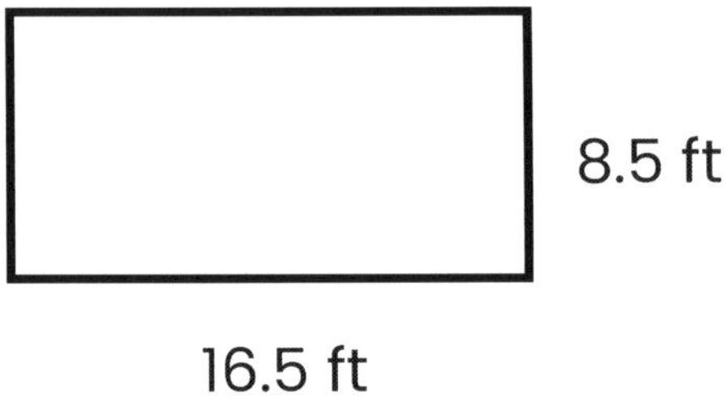

3. Area = ________ in^2

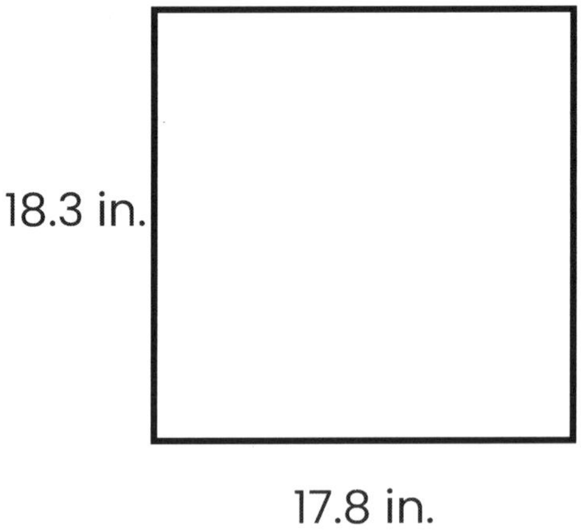

4. Area = ________ m^2

5. A room measures 3.5 meters by 4.2 meters. What is the area of the room in square meters?

Divide each compound figure into rectangles in order to find the total area of the figure.

1. Area = ________ units²

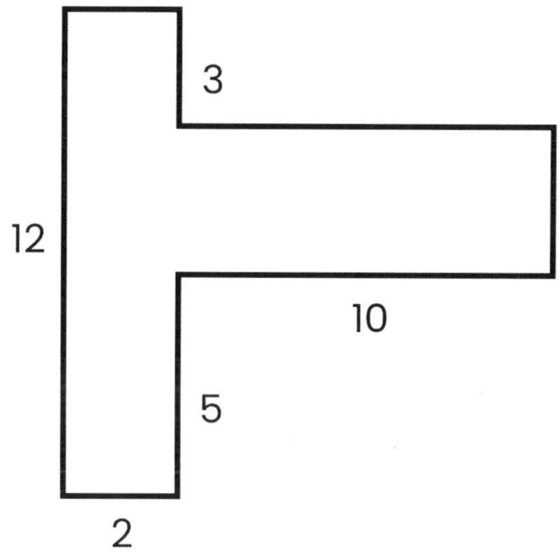

2. Area = ________ units²

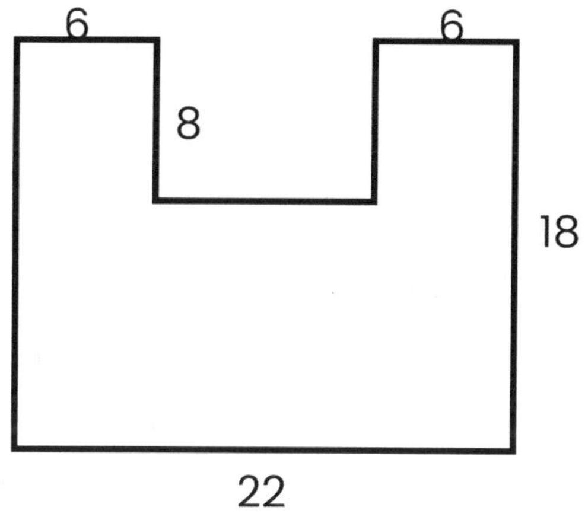

3. Area = ________ units²

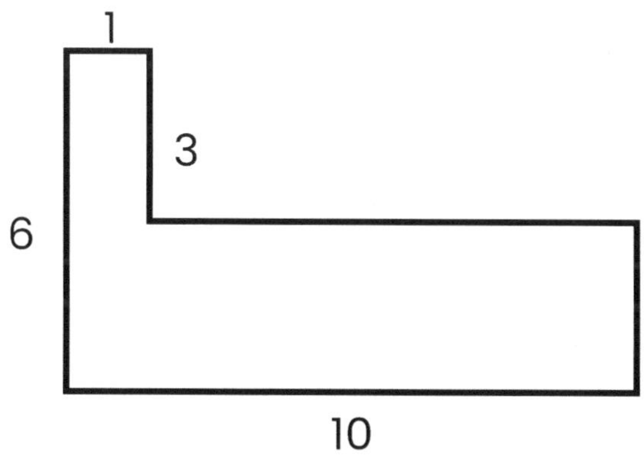

4. Area = ________ units²

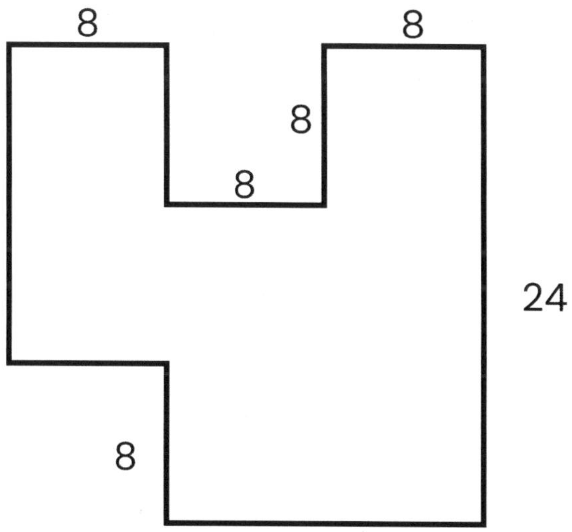

Find the area of the shaded region.

EX Area = 22 units²

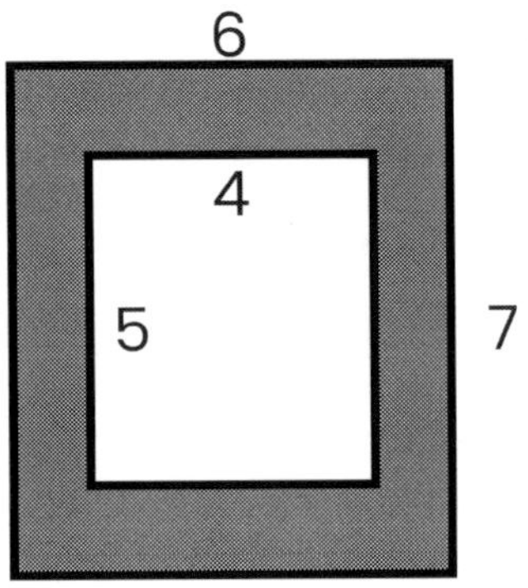

1. Find the area of the outer rectangle.

$6 \times 7 = 42$

2. Find the area of the inner rectangle.

$4 \times 5 = 20$

3. Subtract to find the area of the shaded region.

$42 - 20 = 22$

1 Area = ________ units²

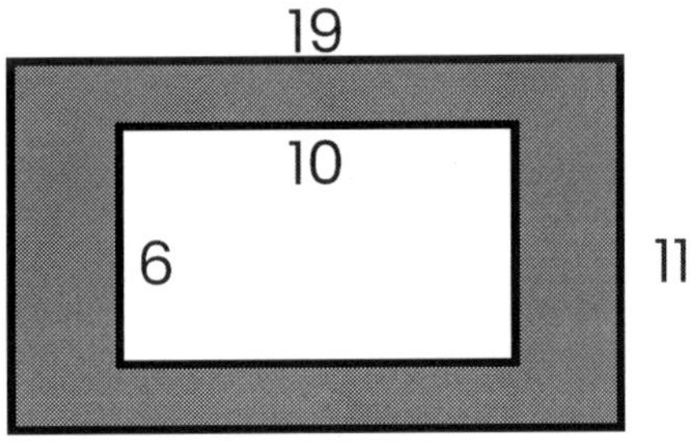

2 Area = ________ units²

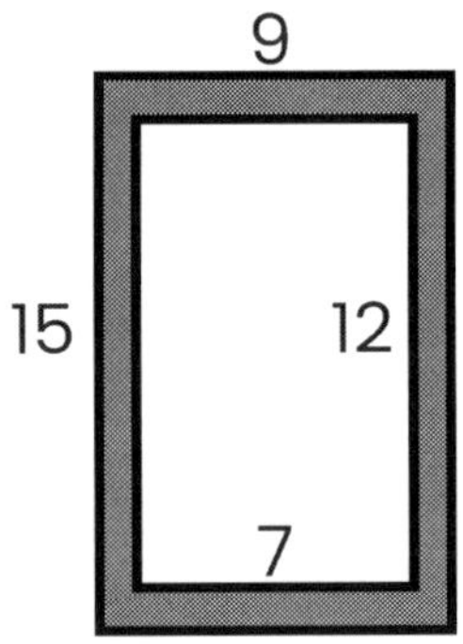

3 Area = ________ units²

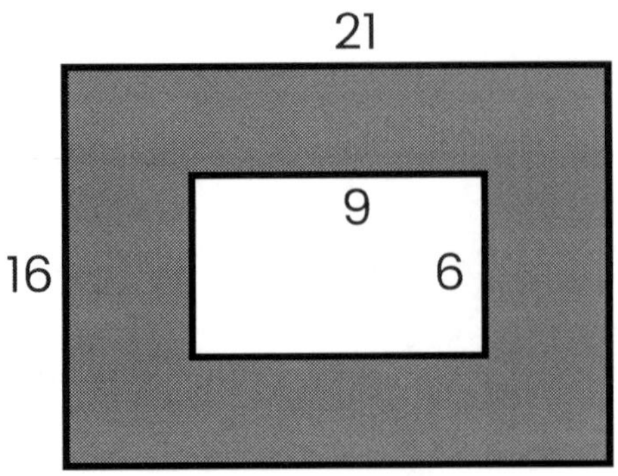

4 Area = ________ units²

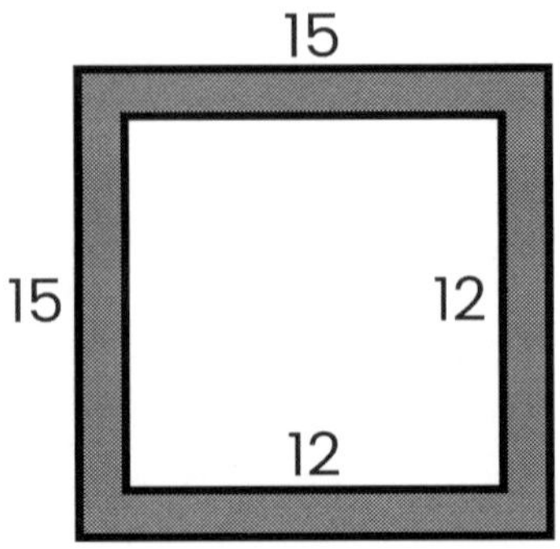

Solve each word problem involving perimeter or area.

1. Peggy painted a 20 inch by 30 inch canvas. She wants to frame it with a wooden border. What will be the length of the border?

2. A television screen measures 38 inches by 21 inches. What is the area of the screen?

3. A certain city block measures 80 m by 250 m. If you walked all the way around the block, what would be the total distance traveled?

4. A window measures 36 inches by 60 inches. How many linear feet of trim would be needed to border the window?

5. A picture frame measuring 11 by 14 inches holds an 8 by 10 inch photo bordered by a mat that fills the rest of the frame. What is the visible area of the mat?

Geometric Measurement

Find the volume of each object by counting unit cubes.

(EX) Volume = 5 units³

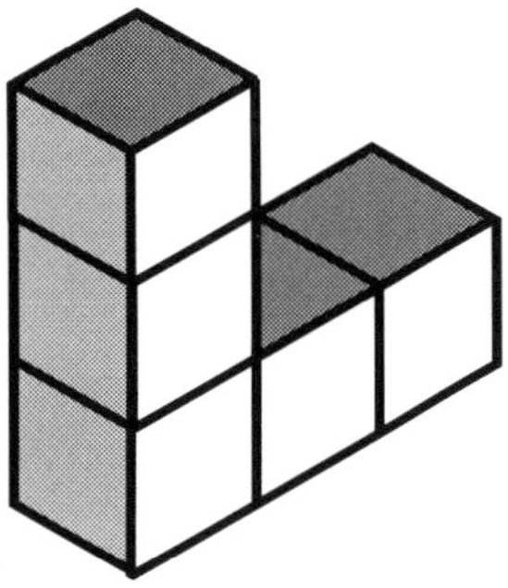

(1) Volume = ________ units³

(2) Volume = ________ units³

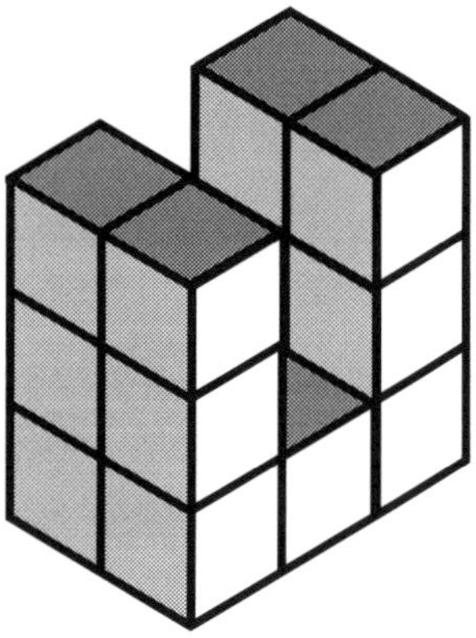

(3) Volume = ________ units³

(4) Volume = ________ units³

(5) Volume = ________ units³

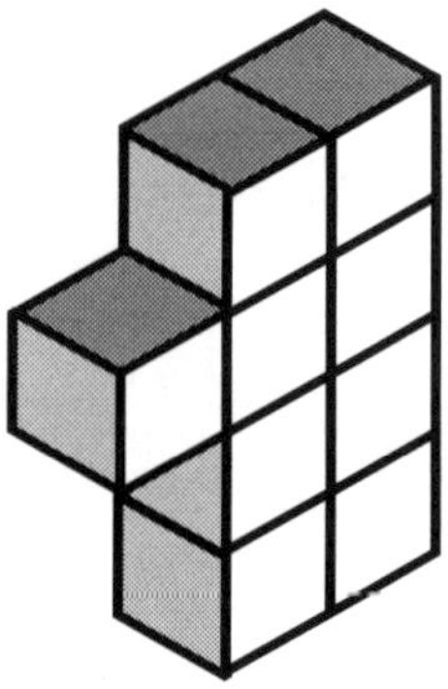

Multiply to find the volume of each rectangular prism.

EX

Length = 4 units

Width = 2 units

Height = 3 units

Volume = 4 × 2 × 3 = 24 units³

1

Length = ______ units

Width = ______ units

Height = ______ units

Volume = ______ × ______ × ______ = ______ units³

2

Length = ______ units

Width = ______ units

Height = ______ units

Volume = ______ × ______ × ______ = ______ units³

3

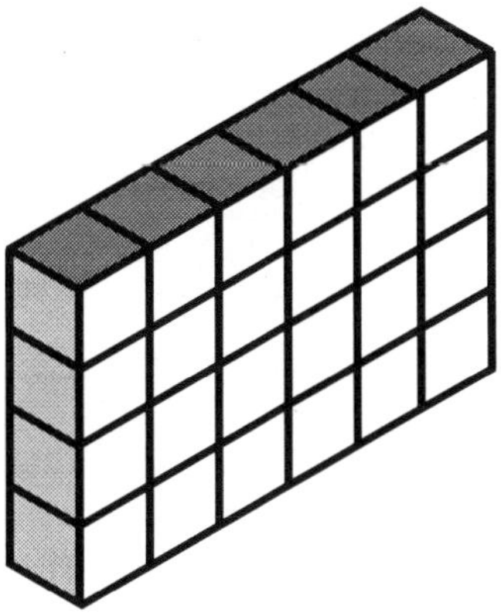

Length = ______ units

Width = ______ units

Height = ______ units

Volume = ______ × ______ × ______ = ______ units³

Geometric Measurement

Find the volume of each rectangular prism.

EX

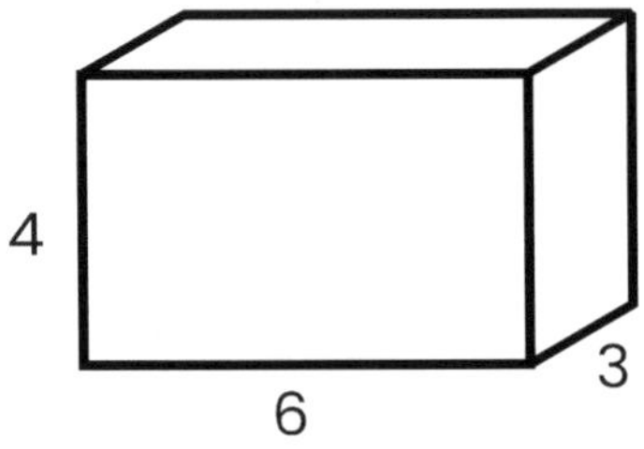

Length = 6 units

Width = 3 units

Height = 4 units

Volume = 6 × 3 × 4 = 72 units³

1

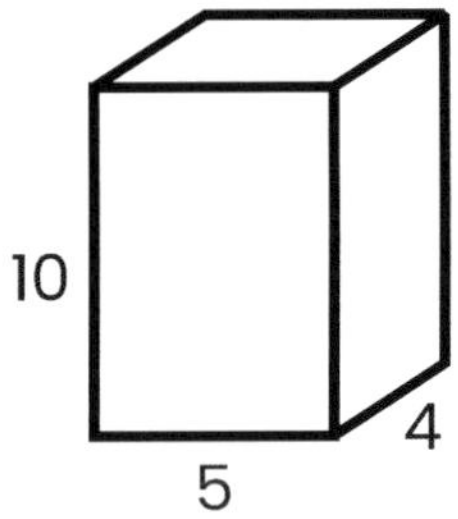

Length = ______ units

Width = ______ units

Height = ______ units

Volume = ______ × ______ × ______ = ______ units³

2

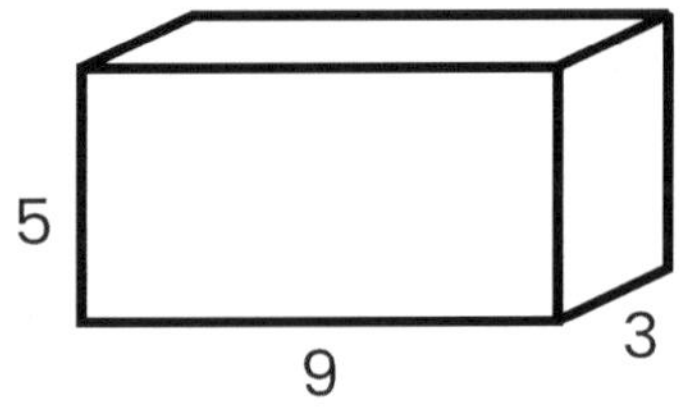

Length = ______ units

Width = ______ units

Height = ______ units

Volume = ______ × ______ × ______ = ______ units³

3

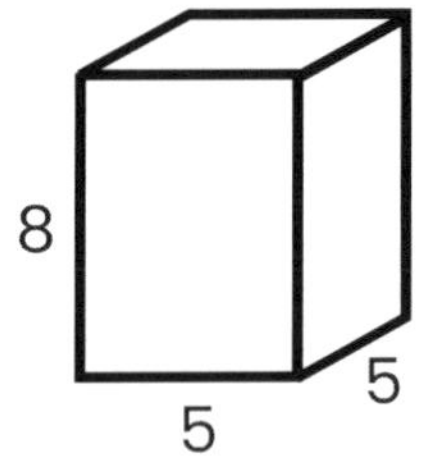

Volume = ______ units³

4

Volume = ______ units³

5

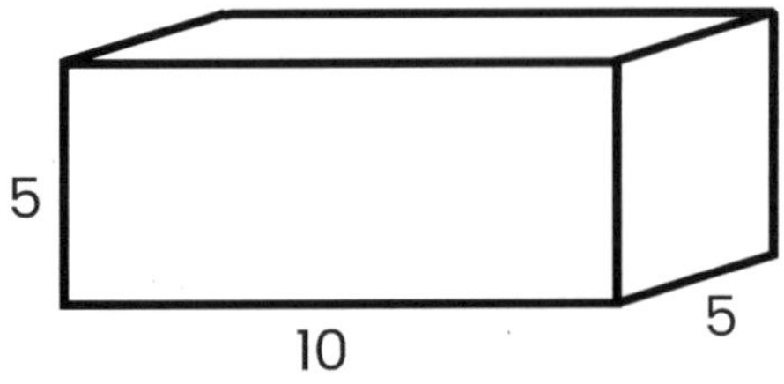

Volume = ______ units³

6

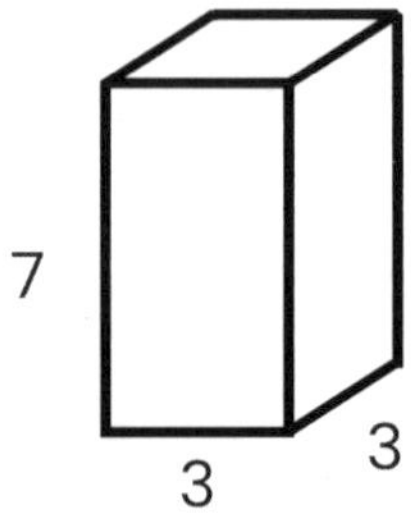

Volume = ______ units³

Find the volume of each cube.

EX

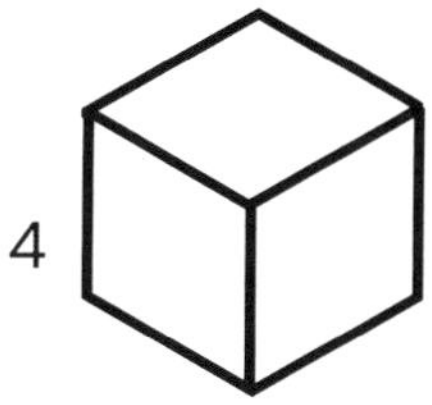

Side = __4__ units

Volume = __4__ × __3__ × __4__ = __64__ $units^3$

1

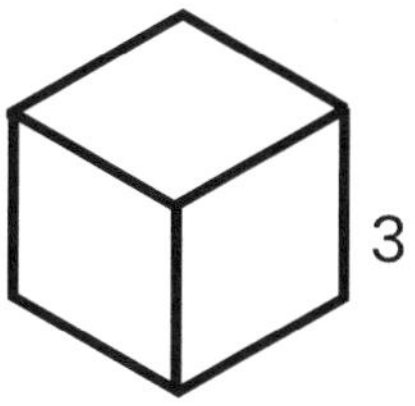

Side = _____ units

Volume = _____ × _____ × _____ = _____ $units^3$

2

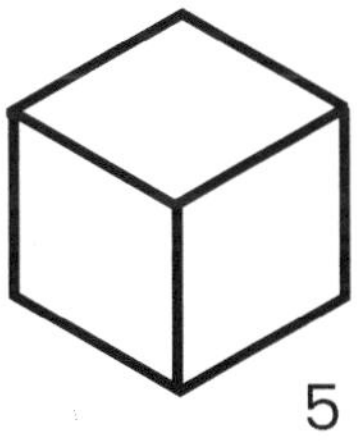

Side = _____ units

Volume = _____ × _____ × _____ = _____ $units^3$

3

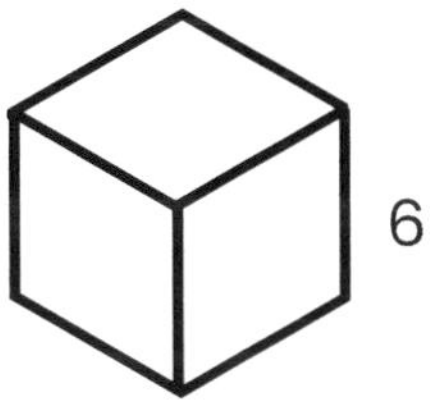

Side = _____ units

Volume = _____ × _____ × _____ = _____ $units^3$

4

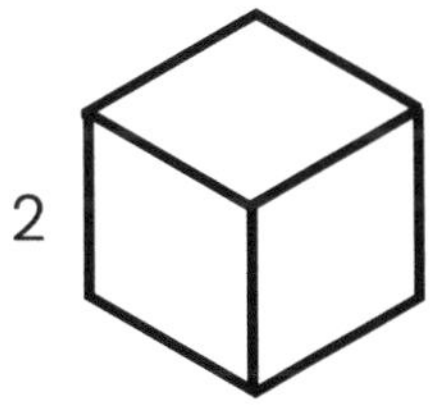

Volume = _____ $units^3$

5

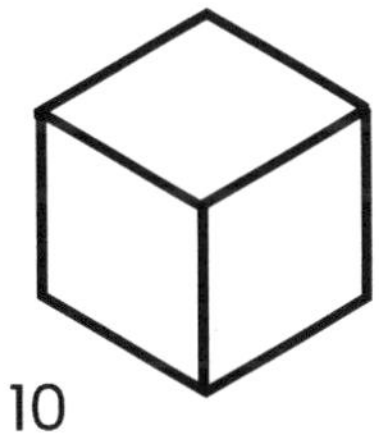

Volume = _____ $units^3$

6 What is the volume of a cube whose sides each measure 7 cm?

Solve each word problem involving volume.

1. A rectangular water tank measures 8 feet long by 5 feet wide by 2 feet tall. What is the volume of the tank?

2. Matt keeps his art supplies in a rectangular storage box that measures 18 cm by 10 cm by 5 cm. What is the volume of the box?

3. Gretchen has a fish tank that is a 20-inch cube. What is the volume of the fish tank?

4. Carson builds a rectangular fort out of blocks that measures 8 inches by 10 inches by 6 inches. What is the volume of the fort?

5. Kate uses cube-shaped containers to store toys. Each one is 12 inches wide. What is the total volume of 2 of these containers?

Fill in the bubble that correctly answers each question.

1. A rectangular prism has a length of 5 cm and a height of 4 cm. If its volume is 160 cm^3, what is its width?

   - O 8 cm
   - O 10 cm
   - O 12 cm

2. Which of these could NOT be the dimensions of a rectangular prism with a volume of 48 m^3?

   - O 4 m × 3 m × 4 m
   - O 2 m × 3 m × 8 m
   - O 4 m × 6 m × 3 m

3. What is the height of a cube whose volume is 512 in^3?

   - O 6 in
   - O 8 in
   - O 9 in

4. Which of these could be the dimensions of a rectangular prism with the same volume as one that measures 3 m × 2 m × 6 m?

   - O 2 m × 5 m × 4 m
   - O 3 m × 3 m × 4 m
   - O 4 m × 2 m × 2 m

5. A cube shaped container 10 inches tall Is filled exactly halfway with water. What volume of the container is filled?

   - O 500 in^3
   - O 750 in^3
   - O 1,000 in^3

Geometric Measurement

Find the volume of each compound figure by breaking it down into rectangular prisms.

EX

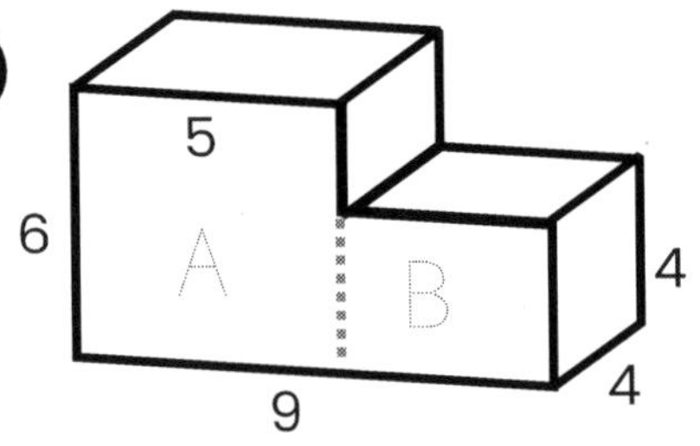

Volume A = 5 × 4 × 6 = 120 units³

Volume B = 4 × 4 × 4 = 64 units³

Total volume= 120 + 64 = 184 units³

1

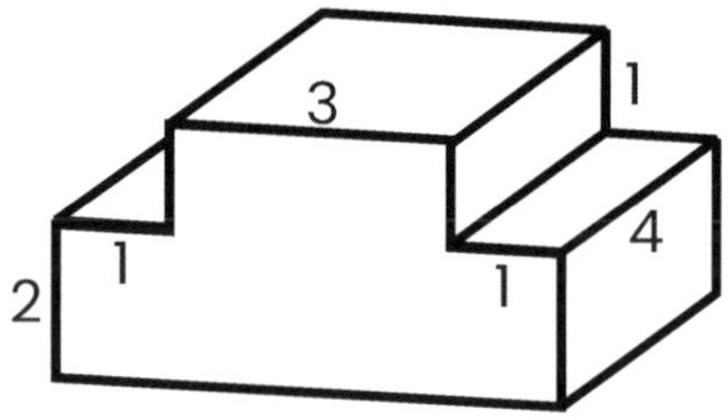

Volume A = _____ × _____ × _____ = _____ units³

Volume B = _____ × _____ × _____ = _____ units³

Total volume= _____ + _____ = _____ units³

2

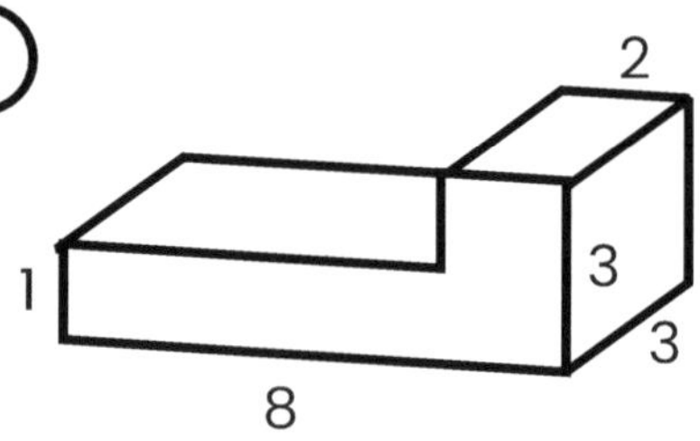

Volume A = _____ × _____ × _____ = _____ units³

Volume B = _____ × _____ × _____ = _____ units³

Total volume= _____ + _____ = _____ units³

3

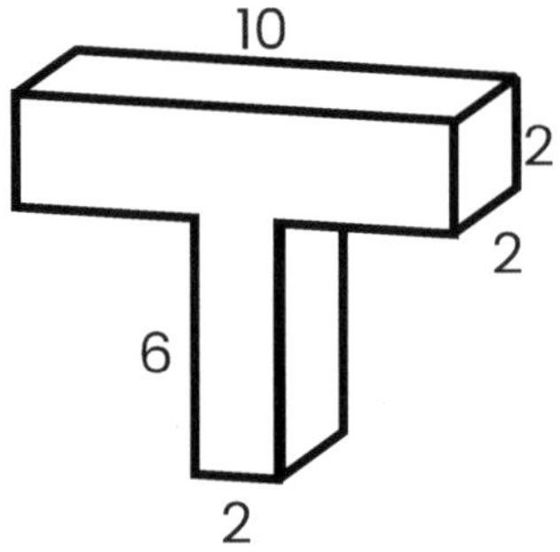

Volume A = _____ × _____ × _____ = _____ units³

Volume B = _____ × _____ × _____ = _____ units³

Total volume= _____ + _____ = _____ units³

Find the volume of each figure.

1

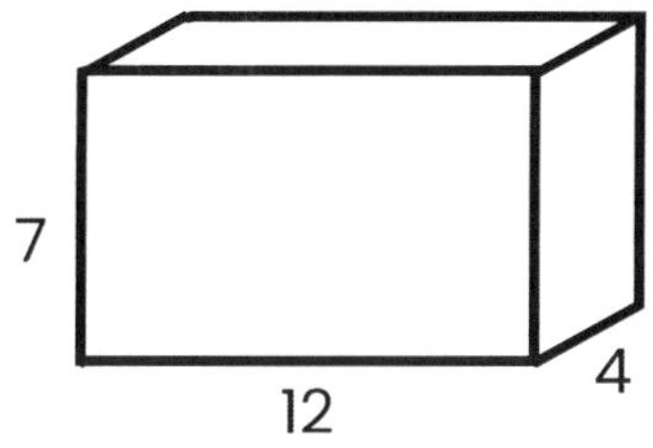

Volume = ______ units³

2

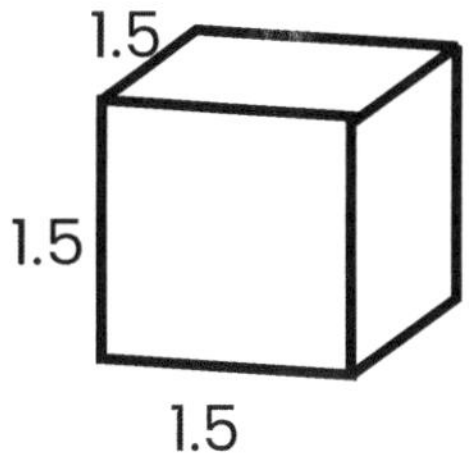

Volume = ______ units³

3

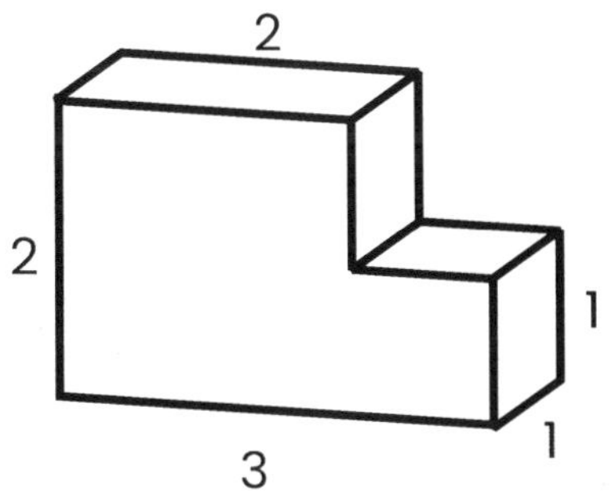

Volume = ______ units³

4

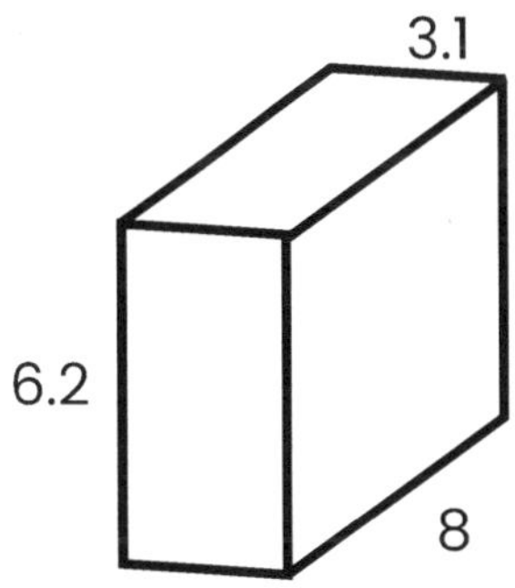

Volume = ______ units³

5

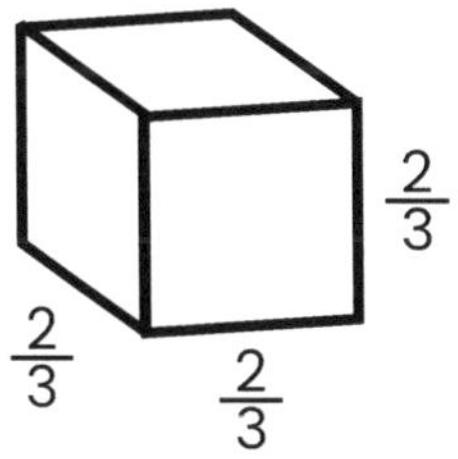

Volume = ______ units³

6

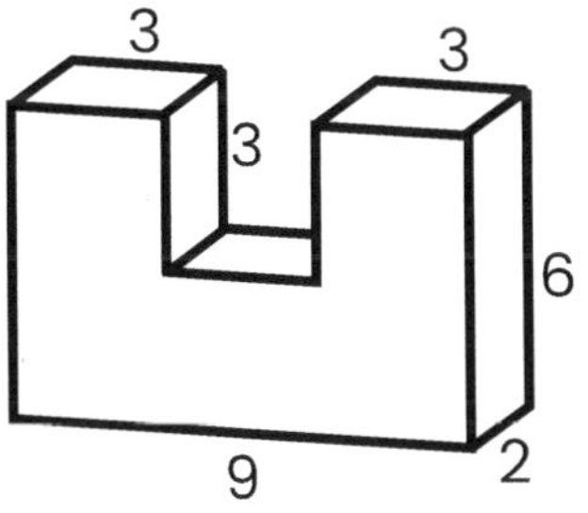

Volume = ______ units³

144 Representing Data

Use the line plot to answer the questions that follow.

The line plot shows the amount of snowfall the same locations received each day in the month of January.

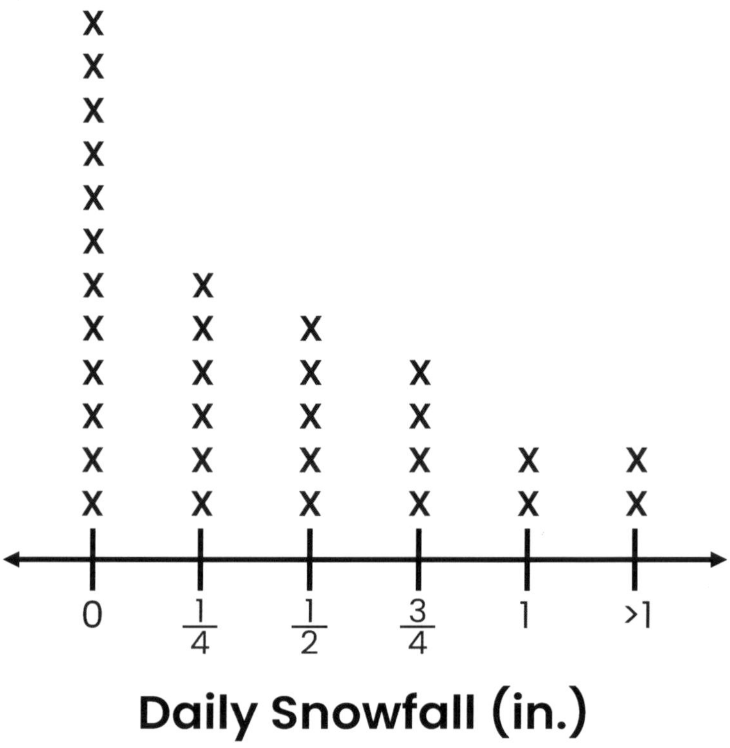

1. On how many days in January did it snow?

2. How many days received less than an inch of snow?

3. How many days received more than half an inch of snow?

4. On January 3, it snowed $\frac{1}{4}$ in. On how many days in January did it snow exactly twice as much as on January 3?

Make a line plot to show the data given. Then, answer the questions below.

Students in a class each planted a radish. Two weeks later, the students measured the heights of their plants to the nearest half-centimeter. The results are recorded in the tally chart.

Height of Plant (cm)	Number of Plants
3.5	|
4.0	|||
4.5	𝍸
5.0	𝍸 |
5.5	|||
6.0	||

Create a line plot to show this data.

1. How many plants were taller than 5 cm?

2. What is difference in heights between the tallest and shortest plants?

3. What was the height of the shortest plant in millimeters?

Use the frequency table below to answer the questions that follow.

The frequency table shows how many 6-hour shifts each of seven coffee shop employees worked last week.

Employee Name	Shifts Worked
Alexis	5
Brendan	4
Cassidy	6
Drake	5
Evelyn	4
Finn	5
Gretchen	3

1. Which employee worked the most shifts last week?

2. How many employees worked at least 5 shifts last week?

3. How many more shifts did Alexis work than Gretchen?

4. How many total hours did Evelyn work last week?

Use the information below to complete the frequency chart. Then, answer the questions that follow.

A board game contains a spinner with sections numbered 1-8. Madeline spun the spinner repeated and recorded the number the spinner landed on each time. Her results are shown below.

2, 4, 1, 2, 5, 7, 6, 8, 1, 5,
4, 4, 5, 8, 7, 1, 3, 8, 4, 6

Use this data to complete the frequency table below to show how often the spinner landed on each number.

Number on Spinner	Frequency
1	
2	
3	
4	
5	
6	
7	
8	

1. How many times did Madeline spin an odd number?

2. How many times did Madeline spin a number greater than 4?

3. How many more times would Madeline have to spin a 3 to equal the number of times she spun a 7?

Use the histogram to answer the questions that follow.

A class held a reading challenge where students recorded the number of minutes they spent reading over the course of two weeks. The results are shown in the histogram below.

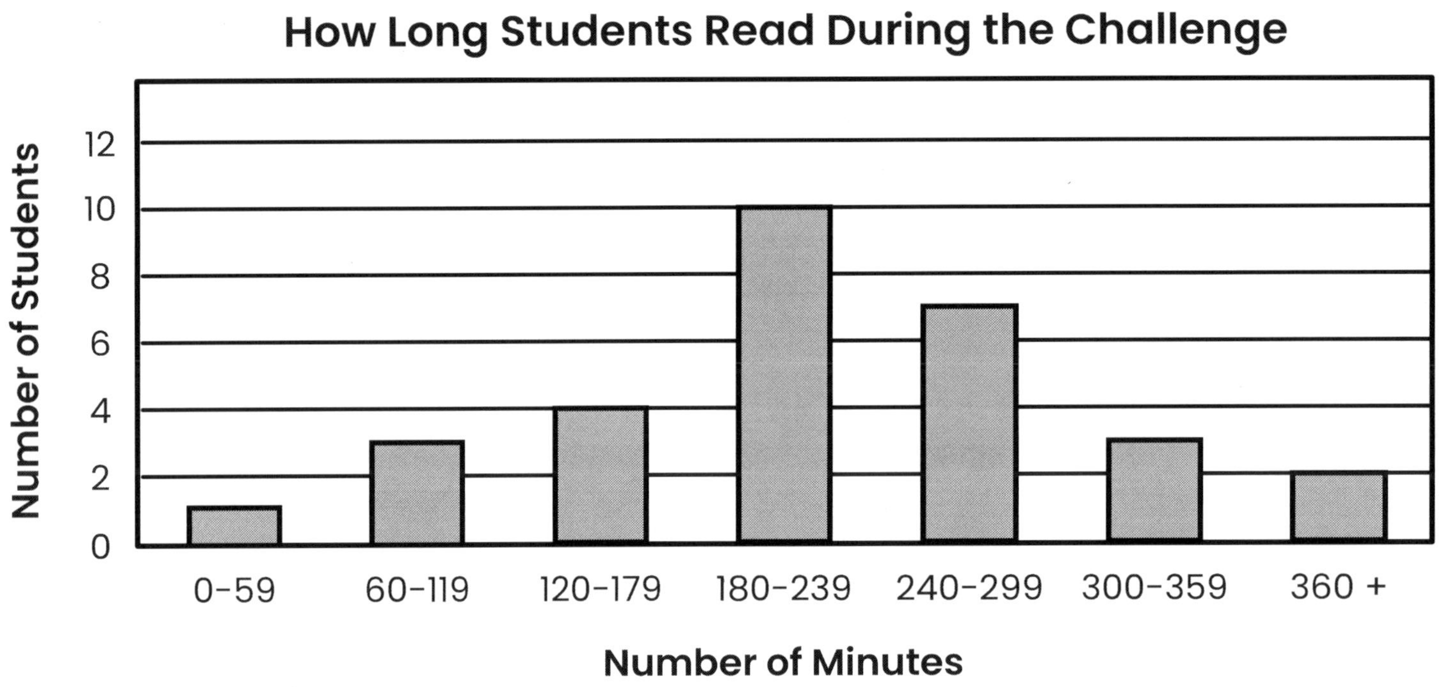

1. How many students read for less than 1 hour?

2. How many students read for at least 300 minutes?

3. What was the most common total number of hours read?

Create a histogram based on the information given.

Students in a class took a 10-question quiz. The number of questions each student got correct out of 10 are recorded on the frequency table. Use this data to complete the histogram below.

Questions Correct	1	2	3	4	5	6	7	8	9	10
Number of Students	0	0	0	1	1	2	4	5	4	3

Quiz Questions Answered Correctly

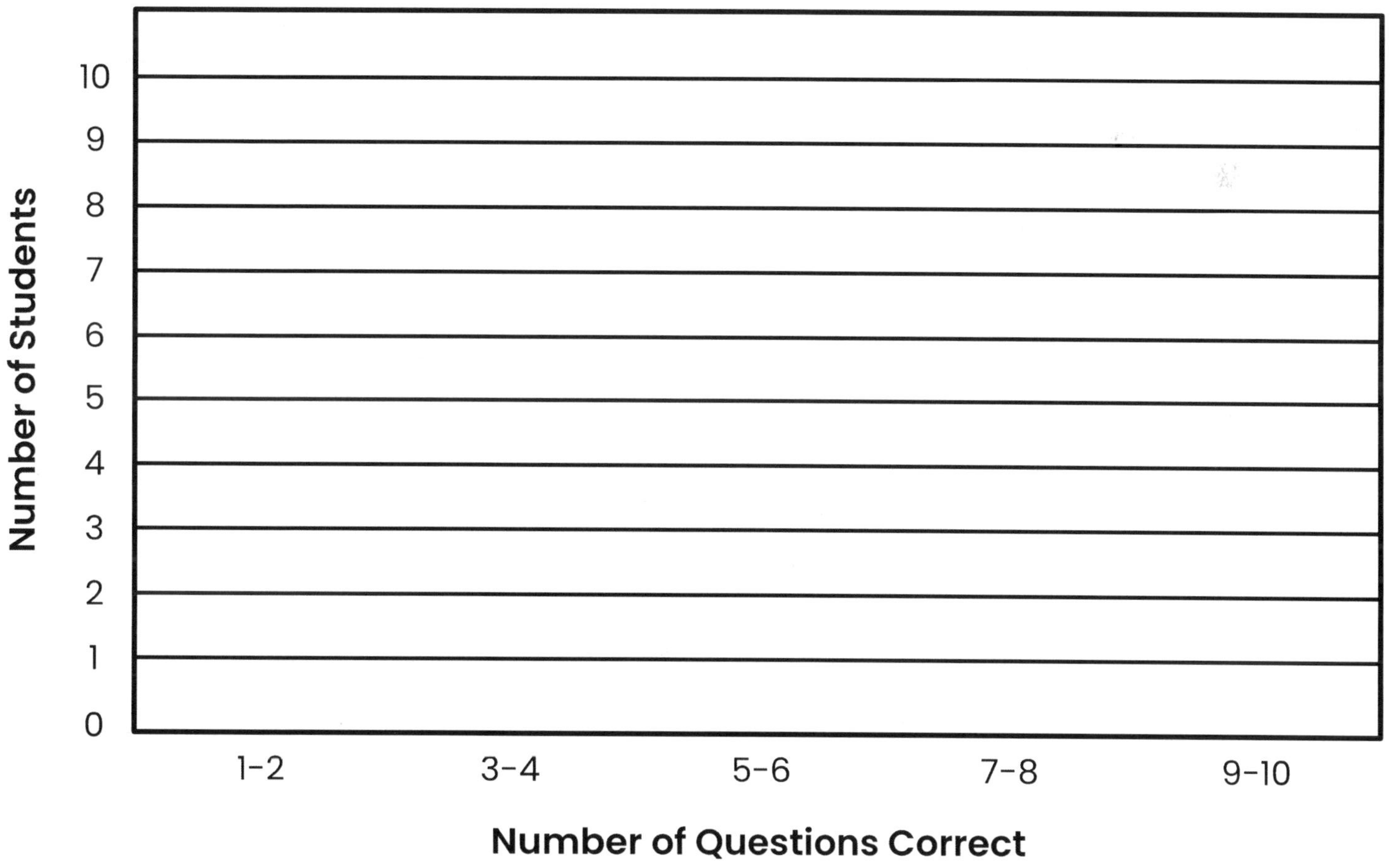

Use the line graph to answer the questions that follow.

Chris kept track of how many days each month he worked out for a year. The results are shown on the graph below.

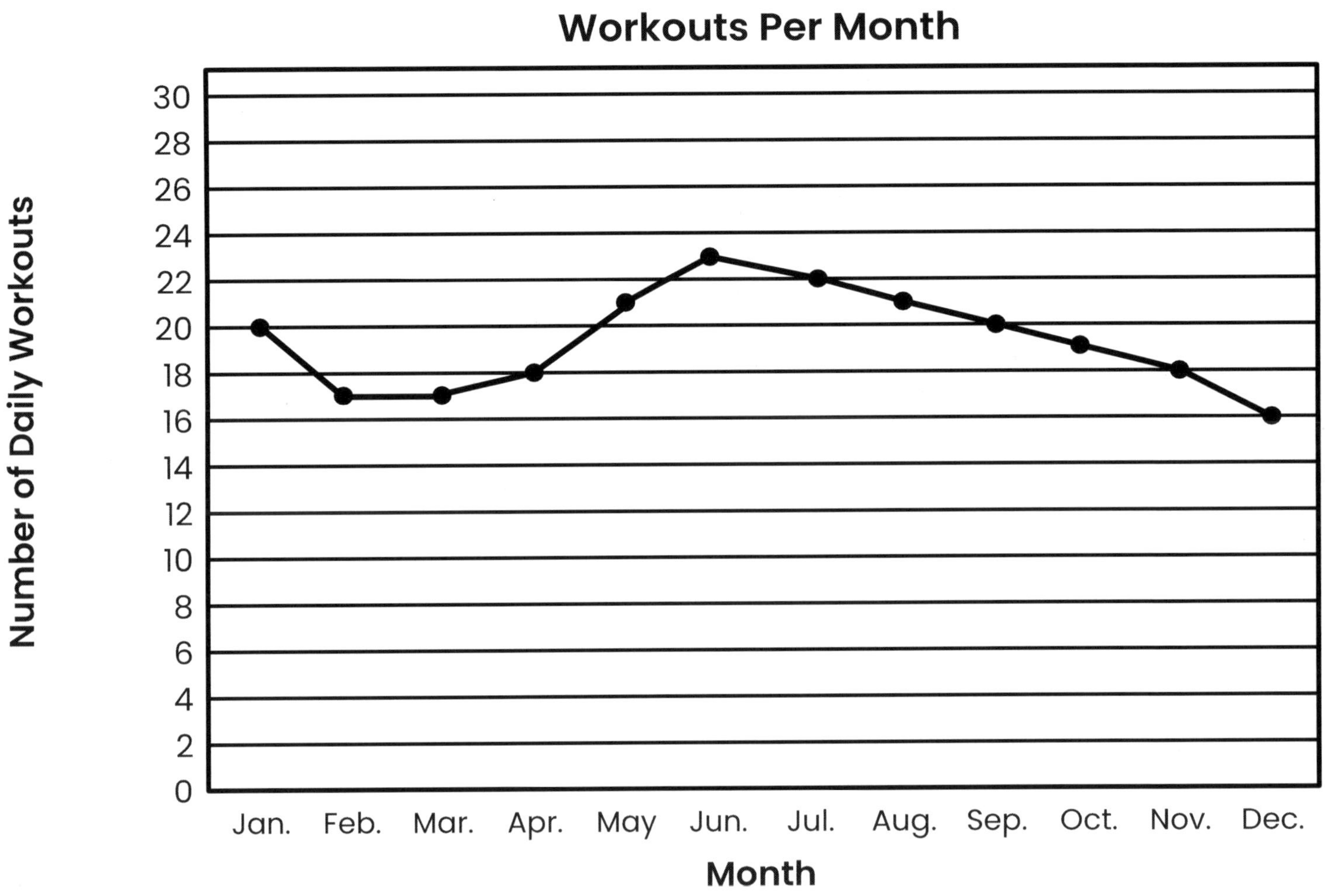

1. How many days did Chris work out from March to May?

2. In how many months did Chris work out more than 20 days?

3. How many more days did Chris work out in July than January?

Create a line graph using the data below.

The table below lists Reagan's height, in inches, each year from ages 2 to 12.

Age	2	3	4	5	6	7	8	9	10	11	12
Height (in)	34	36	40	43	46	49	51	53	55	58	60

Choose an appropriate scale and record the data on the line graph below.

Reagan's Height

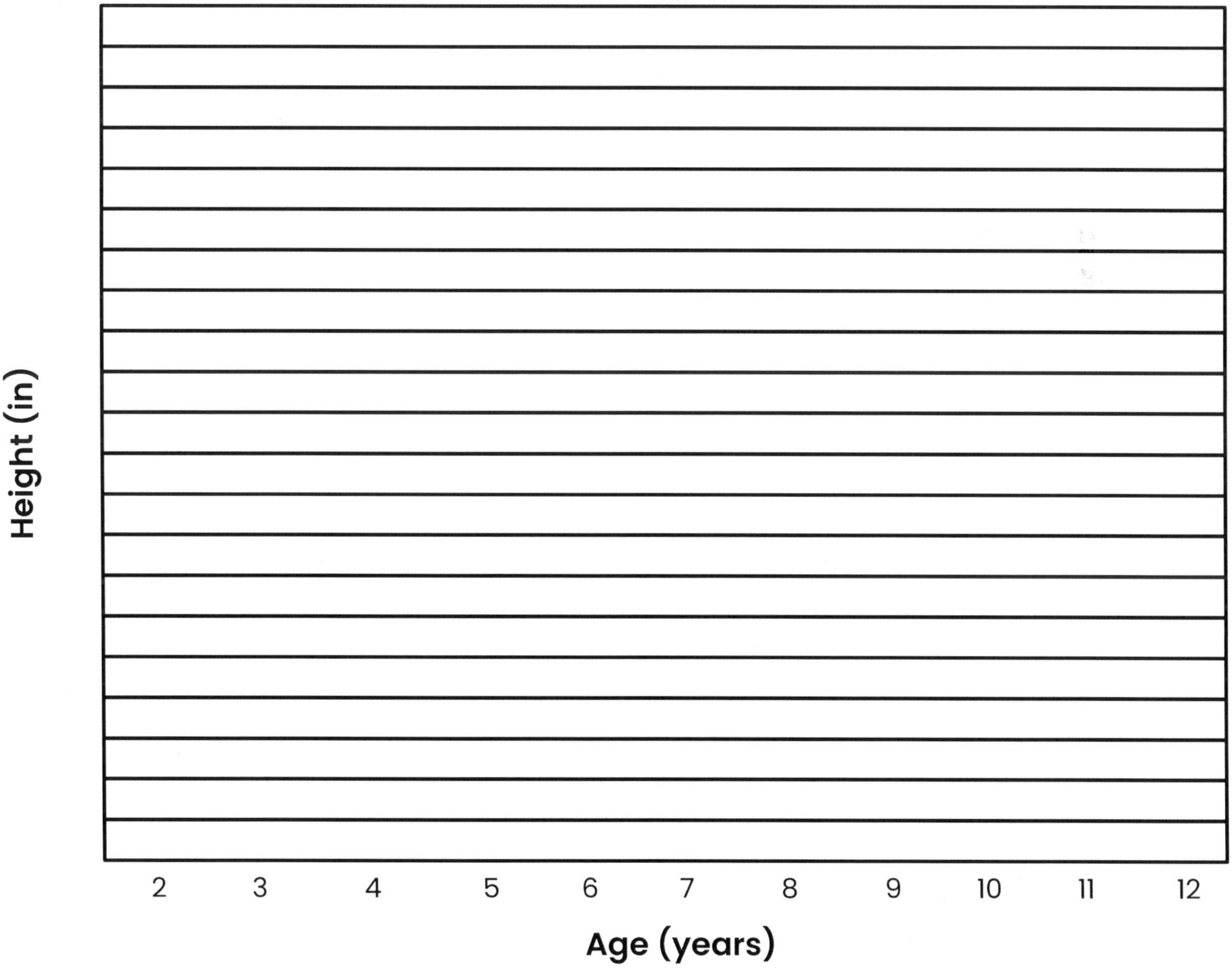

Representing Data

Use the stem and leaf plot to answer the questions that follow.

The stem and leaf plot below shows the number of points scored per game by a basketball team in its 20 most recent games.

Stem	Leaf
5	2 2 8
6	0 0 6 7
7	2 3 4 5 5 5 7 8
8	0 1 2 5 6

1. What was the highest number of points the team scored in a game?

2. How many times did the team score exactly 60 points?

3. In how many games did the team score at least 75 points?

4. In how many games did the team score fewer than 70 points?

5. What was the point difference between the highest and lowest scoring games?

Create a stem and leaf plot using the data below. Then, answer the questions that follow.

Carson measured 10 angles with a protractor. The measures of the angles (in degrees) were 78, 60, 90, 95, 56, 85, 72, 98, 90, and 65. Fill in these measurements on the stem and leaf plot below.

Stem	Leaf

1. What is the difference between the largest and smallest angles?

2. How many angles measured at least 60°?

3. How many of the angles were right angles?

4. Were there more acute, right, or obtuse angles in the group? Explain how you know.

Use the scatter plot to answer the questions that follow.

The scatter plot below shows the average temperature and number of hot chocolates sold each day at a cafe.

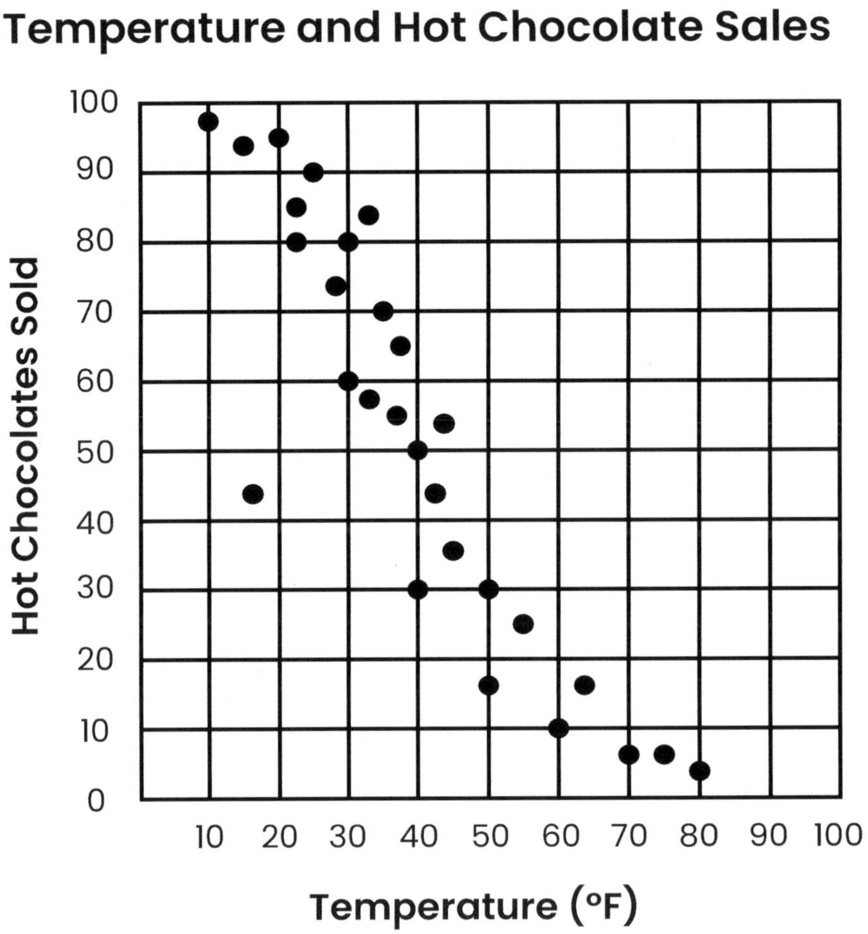

1. Is there a positive correlation, negative correlation, or no correlation between temperature and hot chocolate sales?

2. Draw a line of best fit on the scatter plot.

3. Circle any outliers on the scatter plot.

Create a scatter plot using the data below. Then answer the questions that follow.

The table below lists the heights (in inches) and shoe sizes of 12 women. Record the data as a scatter plot.

Name	Height	Shoe Size
Amy	64	7.5
Brenda	66	8
Claire	67	9
Diana	65	7
Ella	68	9
Fiona	69	9.5
Gina	67	8.5
Helen	65	7.5
Ivy	65	8
Joy	65	8.5
Kate	63	6.5
Lisa	64	7

Women's Shoe Size vs. Height

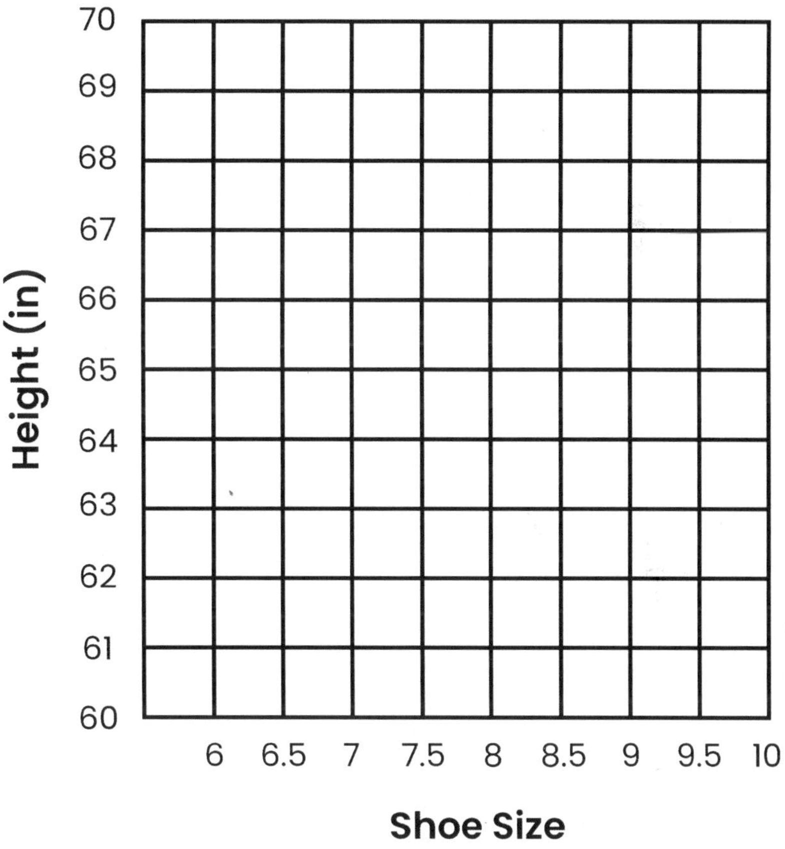

1. Is there a positive correlation, negative correlation, or no correlation between height and shoe size?

2. Molly is 67 inches tall. Based on the trend in the scatter plot, is her shoe size more like to be 7 or 9? Explain.

Coordinate Graphing

Answer the questions about the coordinate plate.

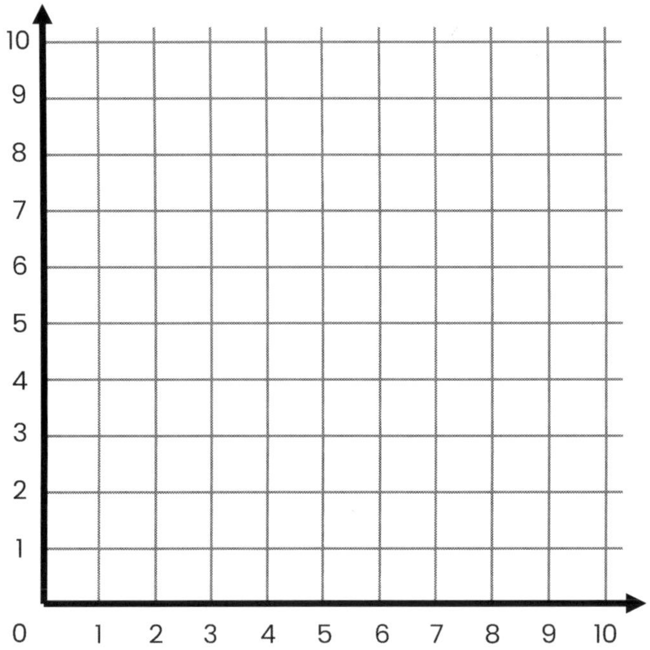

1. Which comes first in an ordered pair: the y-coordinate or the x-coordinate?

2. What are the coordinates of the origin?

3. Is the y-axis horizontal or vertical? Label it on the graph above.

4. Is the x-axis horizontal or vertical? Label it on the graph above.

5. Does the x-coordinate tell you how far to move up/down or left/right from the origin?

Write the coordinates of each point.

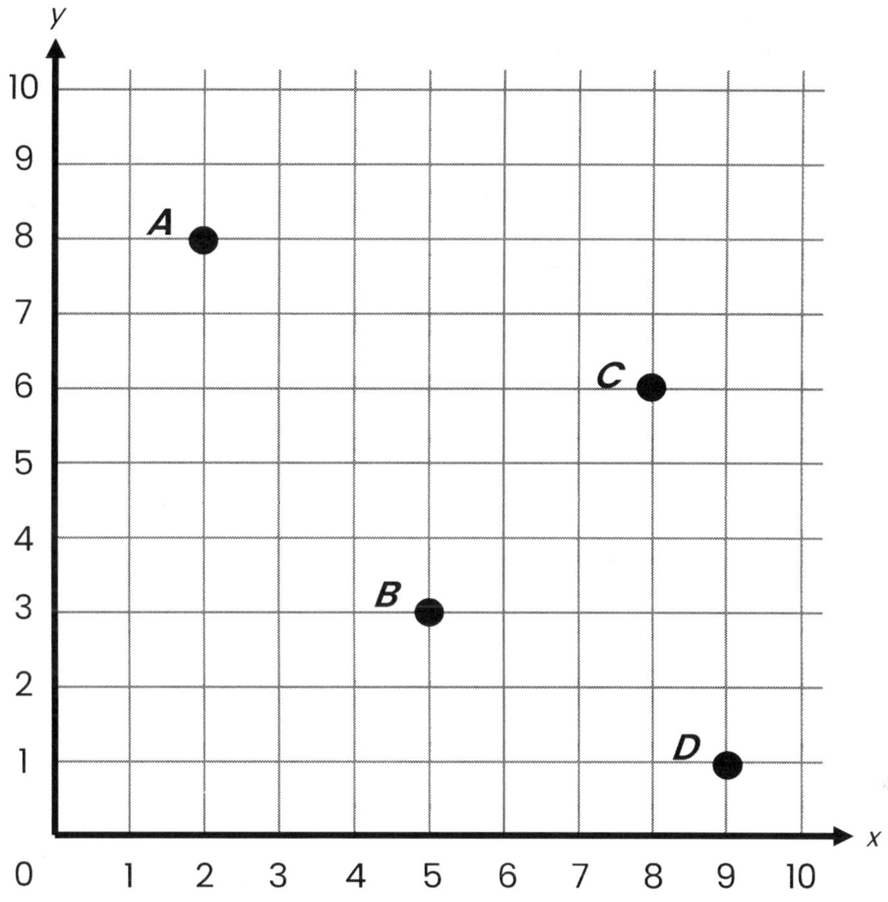

(EX) Point *A*

x-coordinate: 2

y-coordinate: 8

ordered pair: (2, 8)

(1) Point *B*

x-coordinate: ________

y-coordinate: ________

ordered pair: ________

(2) Point *C*

x-coordinate: ________

y-coordinate: ________

ordered pair: ________

(3) Point *D*

x-coordinate: ________

y-coordinate: ________

ordered pair: ________

158 Coordinate Graphing

Graph each point on the coordinate plane.

1. (9, 3)

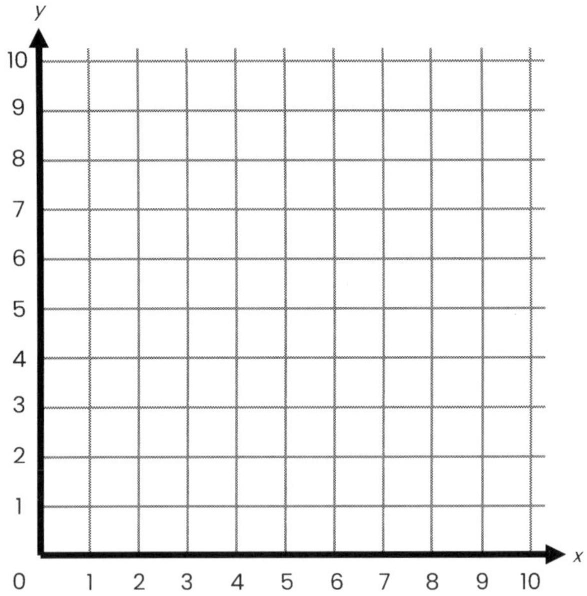

2. (2, 4)

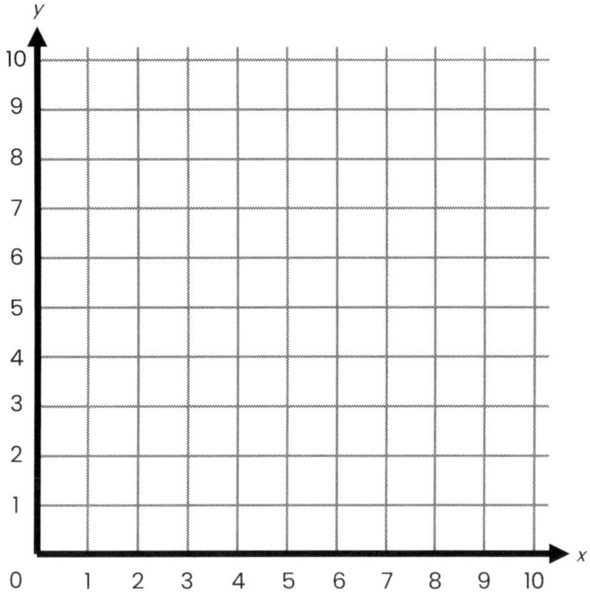

3. (5, 6)

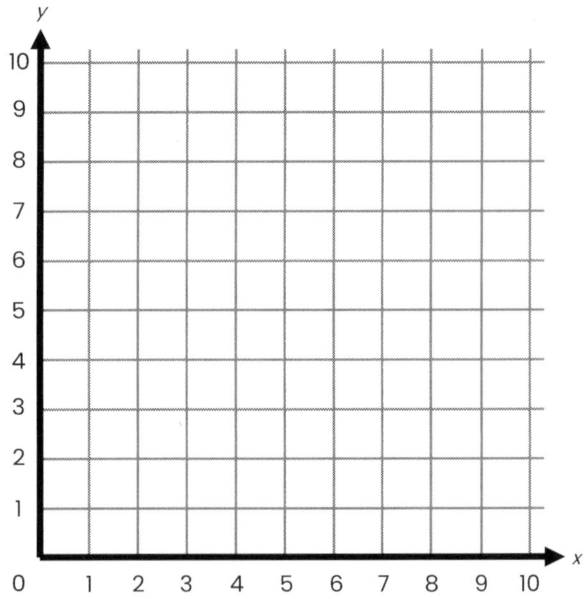

4. (0, 8)

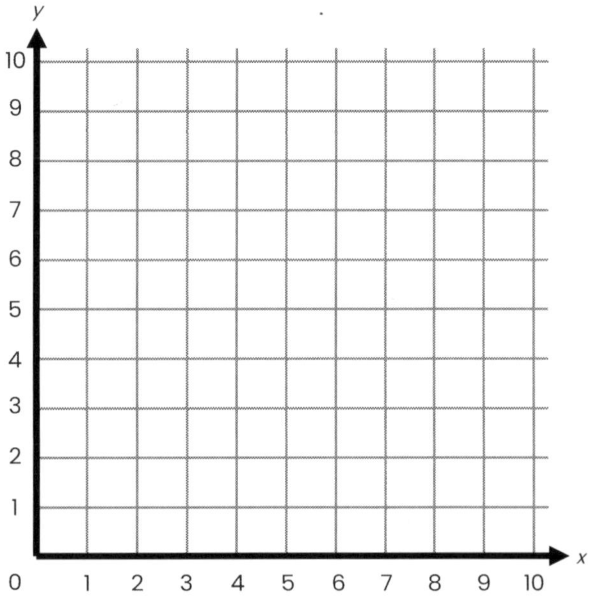

Graph the points from the table on the coordinate plane.

1

x	*y*
4	3
1	7
6	0
2	5
8	8

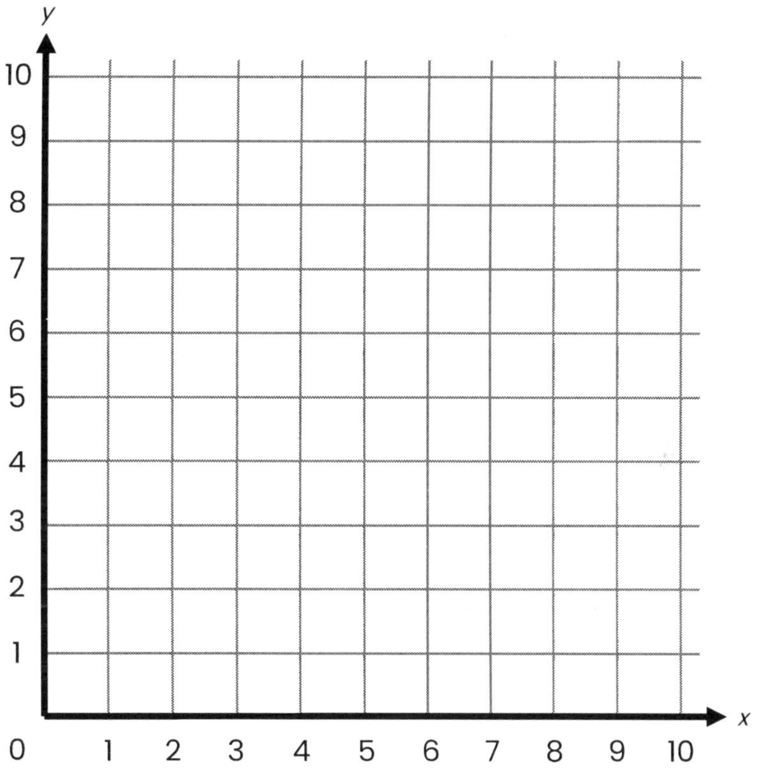

2

x	*y*
7	8
3	9
2	1
0	4
6	3

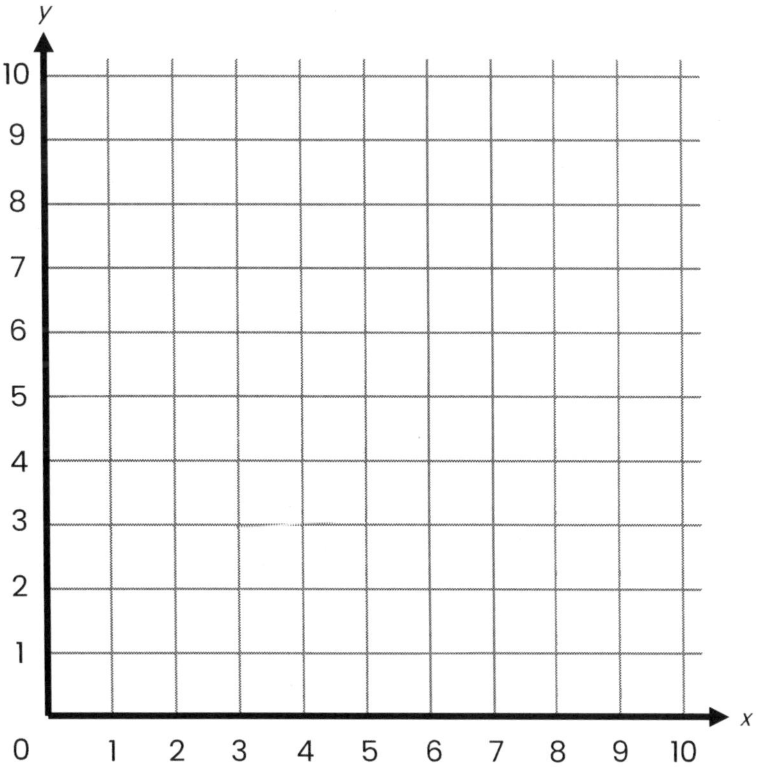

Coordinate Graphing

Complete each input/output table using the rule given. Then, graph the points on the coordinate plane.

① Rule: Add 3.

Input (x)	Output (y)
1	
3	
4	
5	
7	

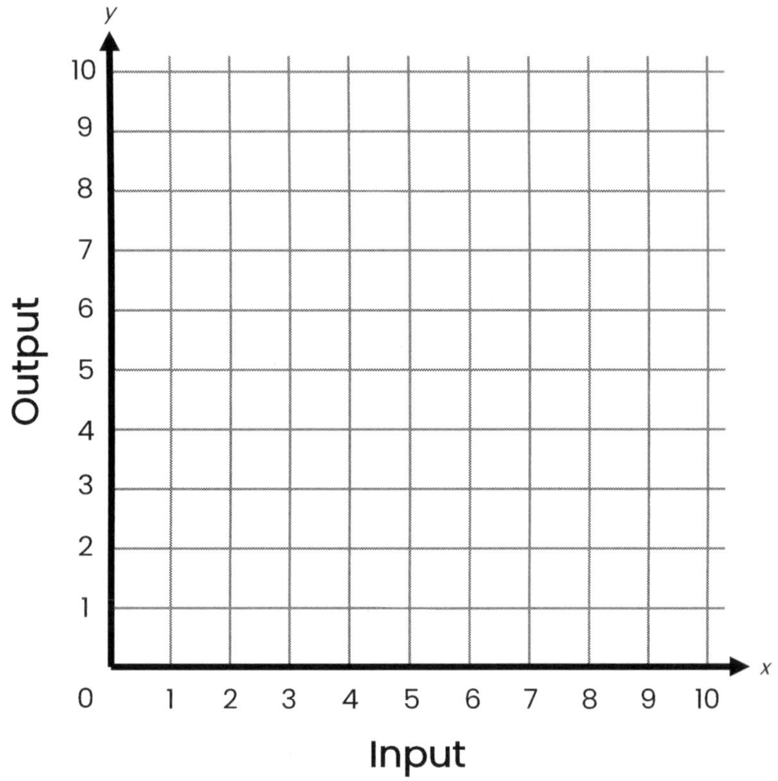

② Rule: Multiply by 2.

Input (x)	Output (y)
1	
2	
3	
4	
5	

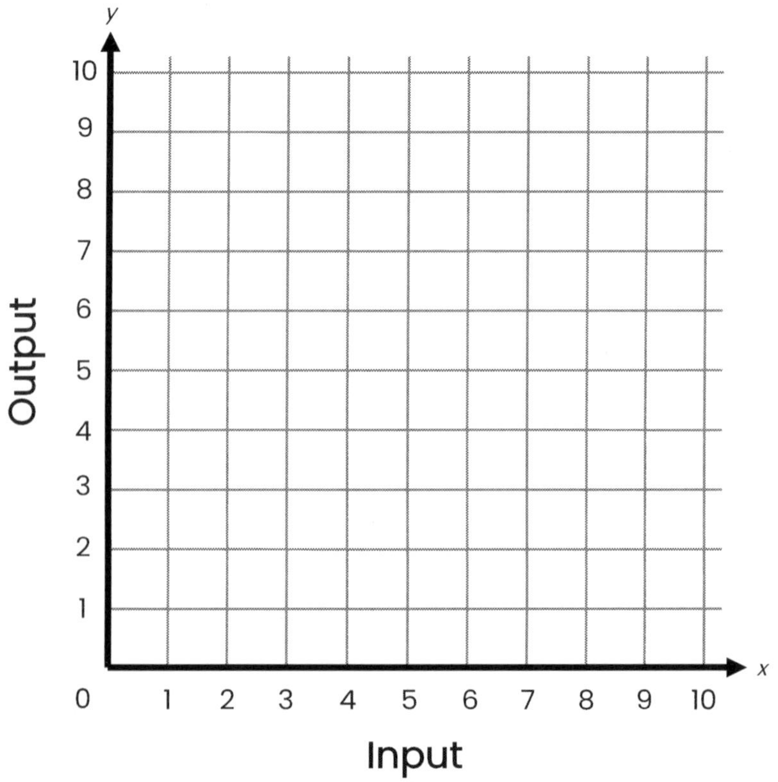

Use the coordinate plane to help you answer each question.

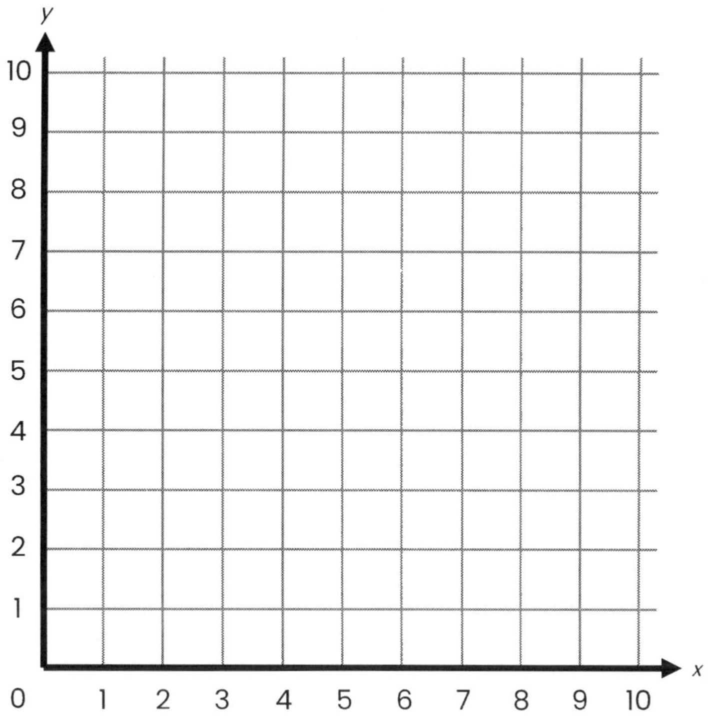

(EX) Start at the point (2, 5). Move 3 units to the right. At what point do you end? (5, 5)

(1) Start at the point (8, 1). Move 2 units up. At what point do you end? ________

(2) Start at the point (4, 4). Move 1 unit to the left. At what point do you end? ________

(3) Start at the point (9, 8). Move 4 units down. At what point do you end? ________

(4) Start at the point (5, 5). Move 1 unit up and 2 units to the left. At what point do you end? ________

(5) Start at the point (7, 6). Move 2 units to the right and 3 units down. At what point do you end? ________

162 Coordinate Graphing

Write the coordinates of the vertices of each shape graphed on the coordinate plane.

EX

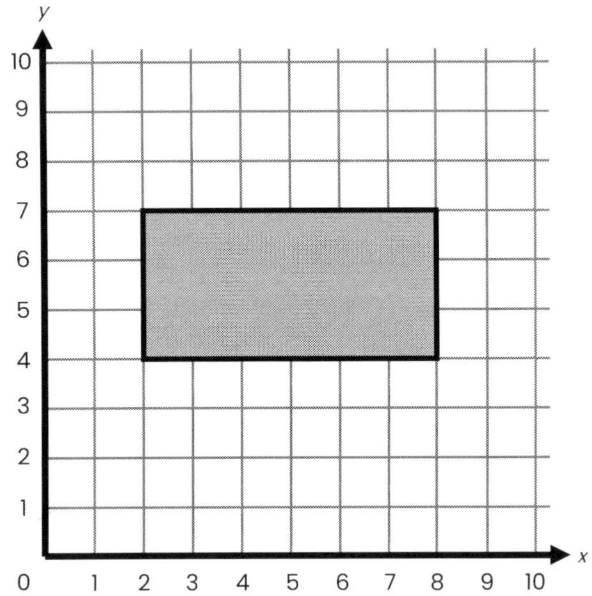

Vertices: (2, 7) ________ ________

________ ________

1

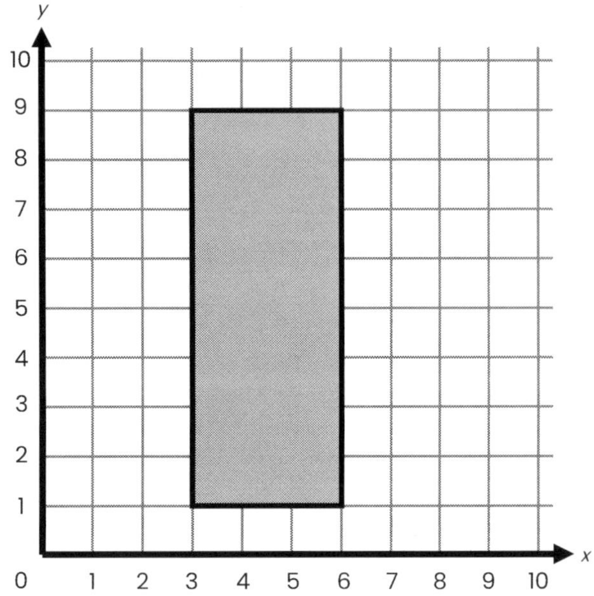

Vertices: ________ ________

________ ________

2

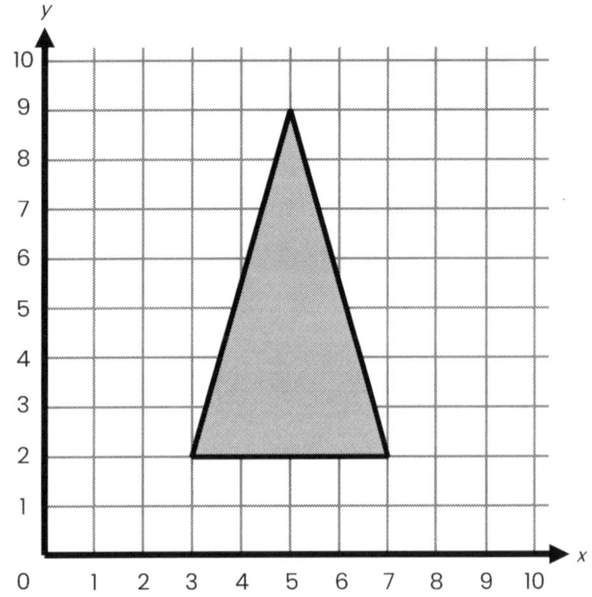

Vertices: ________ ________

____

3

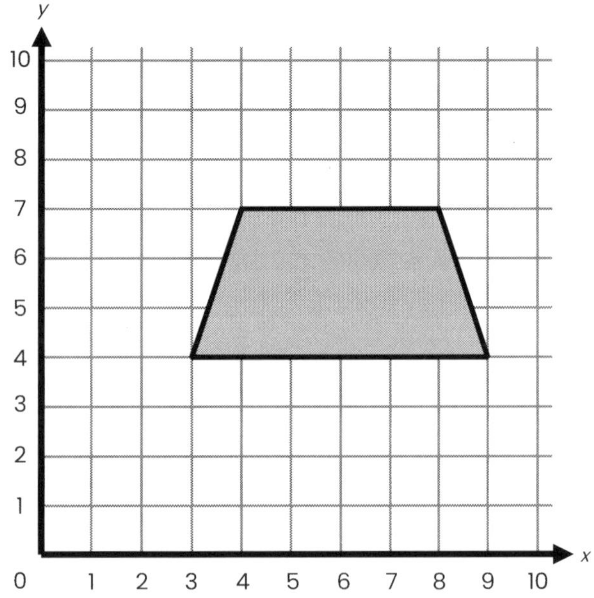

Vertices: ________ ________

________ ________

On each coordinate plane, plot each set of coordinates. Then, connect the points with line segments to draw a shape.

1. Draw a triangle with the vertices (1, 5), (6, 5), and (6, 10).

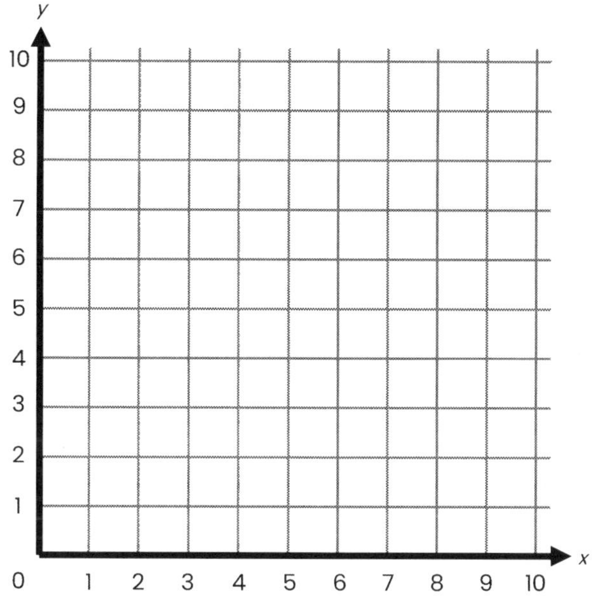

2. Draw a square with the vertices (2, 4), (2, 9), (7, 4), and (7, 9).

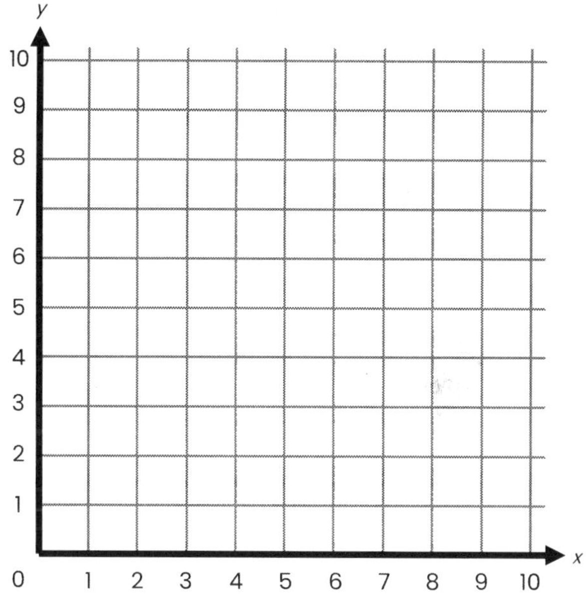

3. Draw a rectangle with the vertices (3, 1), (3, 6), (5, 1), and (5, 6).

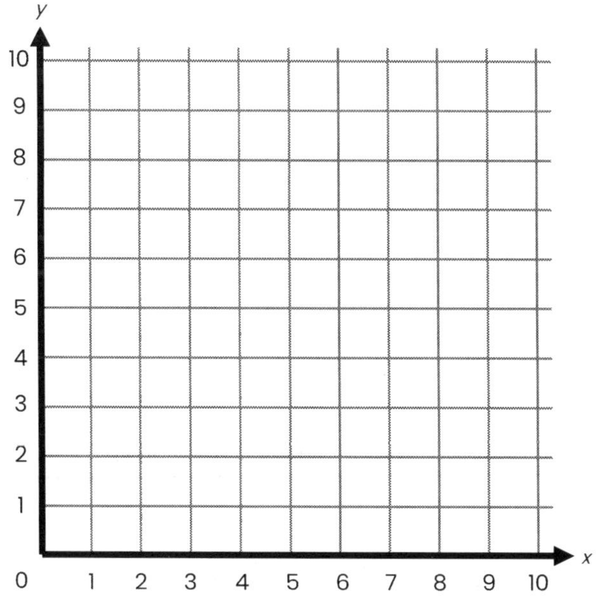

4. Draw a triangle with the vertices (3, 8), (6, 5), and (9, 8).

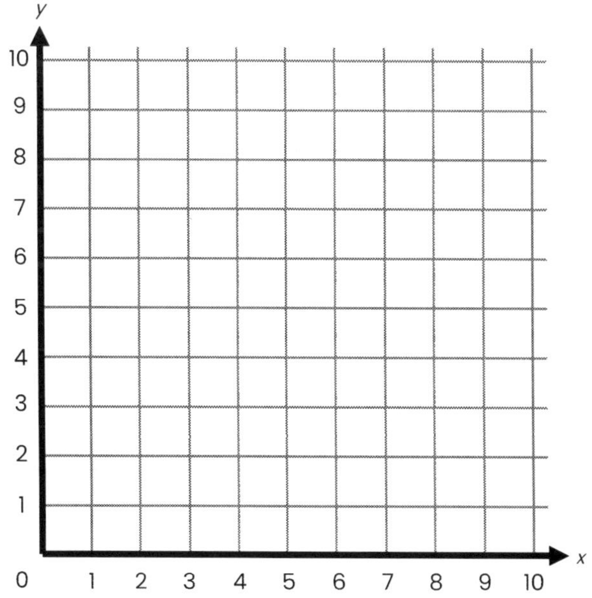

Made in United States
Troutdale, OR
11/03/2026

56845133R00091